HEIDEGGER AND ANIMALS

New Heidegger Research

Series Editors:
Gregory Fried, Professor of Philosophy, Boston College, USA
Richard Polt, Professor of Philosophy, Xavier University, USA

The New Heidegger Research series promotes informed and critical dialogue that breaks new philosophical ground by taking into account the full range of Heidegger's thought, as well as the enduring questions raised by his work.

Titles in the Series

Time and Trauma: Thinking Through Heidegger in the Thirties, Richard Polt
Contexts of Suffering: A Heideggerian Approach to Psychopathology, Kevin Aho
Heidegger's Phenomenology of Perception: An Introduction, Volume I, David Kleinberg-Levin
Confronting Heidegger: A Critical Dialogue on Politics and Philosophy, Edited by Gregory Fried
Proto-Phenomenology, Language Acquisition, Orality and Literacy: Dwelling in Speech II, Lawrence J. Hatab
Transcending Reason: Heidegger on Rationality, Edited by Matthew Burch and Irene McMullin
The Fate of Phenomenology: Heidegger's Legacy, William McNeill
Agency, Freedom, and Responsibility in the Early Heidegger, Hans Pedersen
Heidegger's Phenomenology of Perception: Learning to See and Hear Hermeneutically, Volume II, David Kleinberg-Levin
Towards a Polemical Ethics: Between Heidegger and Plato, Gregory Fried
Thought Poems: A Translation of Heidegger's Verse, Martin Heidegger—Translated by Eoghan Walls
Correspondence: 1919–1973, Martin Heidegger and Karl Löwith—Translated by J. Goesser Assaiante and S. Montgomery Ewegen
Heidegger in the Literary World, Edited by Florian Grosser and Nassima Sahraoui
Heidegger and Music, Edited by Casey Rentmeester and Jeff R. Warren
Heidegger and the Holy, Edited by Richard Capobianco
The Politics of Attention and the Promise of Mindfulness, Lawrence Berger

Heidegger, Dasein, and Gender: Thinking the Unthought,
Edited by Patricia Glazebrook and Susanne Claxton
Heidegger's Being and Time: Paraphrased and Annotated, Volume 1,
Thomas Sheehan
Heidegger's Russians: Bibikhin and Dugin on the Other Beginning,
Jeff Love and Michael Meng
Irish Phenomenology: Thinking with Heidegger and Beyond,
Edited by Róisín Lally and Dan O'Dea Bradley
The Mystery of Being: Reconsidering the Mystical Element in Heidegger's Thought,
Edited by Erik Kuravsky
Heidegger and Parmenides, Edited by Laurence Hemming and Aaron Turner

HEIDEGGER AND ANIMALS

New Insights from Ethology and Evolutionary Biology

Jerome C. Regier

BLOOMSBURY ACADEMIC
LONDON • NEW YORK • OXFORD • NEW DELHI • SYDNEY

BLOOMSBURY ACADEMIC
Bloomsbury Publishing Plc, 50 Bedford Square, London, WC1B 3DP, UK
Bloomsbury Publishing Inc, 1359 Broadway, New York, NY 10018, USA
Bloomsbury Publishing Ireland, 29 Earlsfort Terrace, Dublin 2, D02 AY28, Ireland

First published in Great Britain 2026

Cover design by Neil Cotterill
Cover image © disk/ Getty Images

A catalogue record for this book is available from the British Library.

A catalog record for this book is available from the Library of Congress.

ISBN: HB: 978-1-350-60413-1
ePDF: 978-1-350-60414-8
eBook: 978-1-350-60415-5

Series: New Heidegger Research

Typeset by Integra Software Services Pvt. Ltd.

CONTENTS

TABLES

Lists

PREFACE

I have been curious about our fellow animals[1] as far back as I have memories, and their responses to me have only amplified my curiosity. Thus, I was sad and confused when I eventually became aware of their dire situation in our increasingly human- and human-technology-dominated world (Mindell 2024; Rus and Mone 2024). I asked myself: Why is this situation tolerable to humanity? Why have our fellow animals been victimized through human-caused habitat degradation and displacement, leading to massive loss of lives? Even those animals to whom we humans are directly responsible have been victimized. Consider especially the scale of animal abuse in laboratory testing and experimentation, in the "exotic" pet trade, and in industrialized food production. All this is well-known and, as a result, the claim that we are living in the midst of a human-caused sixth mass extinction can no longer be dismissed as hyperbole. But is this a problem and, if so, what kind of problem is it, and for whom, and why is it a problem? Alas, despite the readiness with which many people would respond, answers are multiple and contradictory. Why is this the case? And why is the very technological thinking that has guided this disaster also frequently seen as the means by which to mitigate it, as though technological thinking itself cannot be a problem? Why are our fellow animals oftentimes invisible to us, for example, when we voice concern about *species* extinctions? Why has the crisis confronting our fellow animals become evident to many only now that lifestyles of wealthy, industrialized, and powerful people and countries are being threatened by global climate change? People generally agree that understanding this problem—even before considering solutions—cannot be confined to making careful observations about the natural world because human interpretations of that world require more than the facts themselves. For example, people may care more about their own economic well-being than the lives of others, particularly when those others are different, like our fellow animals. At other times, people are unwilling to challenge their traditional ways of living, even when discordances appear—like mass extinction (of others!). And frequently, people are quickly satisfied by offering incremental solutions (e.g., larger cages for chickens [see Nicol 2015], "safe" corridors for migrating animals, plant-based meat alternatives) without giving explicit or adequate consideration to what underlies the problem itself. Of course—one might say—the problem is human caused, but still, are we interested in what is it about ourselves, our fellow animals, and the world that has led to this situation? And is change possible, or even desirable?

At this point, some thoughtful people would respond that what is called for is a moral/ethical critique around which political actors could then consolidate. Philosopher Christine Korsgaard's (2018) *Fellow Creatures: Our Obligations to the Other Animals* provides an exemplary argument for what is at stake in our failure to acknowledge the obligations which we ought to have to our fellow animals. Of particular interest to me is Korsgaard's approach in using the writings of another philosopher, Kant (but also Aristotle), as a framework for explicating her view of animals. This approach need not be seen as unwarrantedly esoteric, since philosophers are trained to contemplate fundamental matters, and one such matter is how we relate, and ought to relate, to our fellow animals. Kant claims that how we treat our fellow animals should be largely evaluated by its effect on us humans—a still not uncommon position, I venture to say—but Korsgaard shows through her more animal-friendly eyes that Kant fails to follow the implications of his other thoughts in claiming this. Forty-three years prior to Korsgaard, Peter Singer invigorated the modern field of animal ethics and downstream political activism with his seminal book entitled *Animal Liberation* (more recent editions: Singer 2002, 2023). This too is a compelling philosophical analysis of the treatment of our fellow animals, now within a utilitarian consequentialist framework that aims to find principles which maximize overall well-being. Before, in-between, and after these two publications are numerous others that raise the problematic nature of our relationship to our fellow animals (before: see Clark 2011; Garrett 2011; in-between: see Regan 2004, and Beauchamp and Frey 2011; after: see Nussbaum 2023, which ventures further into political issues).

But what I find striking is that during the reception of the many thoughtful and personally motivating philosophical studies, the situation for our fellow animals continues to deteriorate. We might ask whether this is a personal failing (e.g., people's preferences for their own comfort lead to disregarding reasoned arguments), a social and political failing (e.g., governments are unwilling to support the interests of beings who neither vote nor pay taxes), or whether the ethical analyses themselves are rejected or ignored as such. As regards this last possibility, both Korsgaard and Singer are supportive of much better treatment of animals, and they both argue that an animal's agency and embodiment are central to their own thinking. Another animal-friendly philosopher—Peter Godfrey-Smith (Godfrey-Smith 2020)—places greater emphasis on animal sentience. By contrast, some Christian groups claim that our fellow animals do not have agency or sentience and that the body is of lesser importance than the (disembodied) soul, which animals also do not have. Or alternatively, animal well-being may be considered secondary to human well-being for perceived economic reasons. Clearly, there are multiple ways of thinking about how we ought to relate to our fellow animals, only some of which emphasize the importance of animal well-being.

None of this is to argue that ethics is not a proper forum for discussing the animal problem. Of course, one could argue that ethics is necessarily *human* ethics, since our fellow animals are most often assumed to be incapable of thinking ethically, or even thinking at all. Such claims make a strong distinction between an agential subject—the human—and an object—our fellow animal—thereby fostering the idea that being an animal is fundamentally different from being a human. Like the embodiment issue, the subject-object distinction is not a directly ethical matter.

Perhaps both are more closely related to ontological thinking, that is, to thinking about what entities "are" through their very being. (My understanding of "being" and related words is central to the entire book, but—for now—that desideratum is held in abeyance.) Might it be that the common understanding of animal *being* (as opposed to human *being*) is such that animal-friendly claims become harder for humans to conceive? To address this, we need to explicate an understanding of animal being. That is a task that I—a formally trained biologist with no formal philosophical training—have set for myself. While biology has informed my thinking while writing this book, I see myself largely as doing philosophy. It is a risky proposition for me, but one that I felt compelled to undertake given a nagging feeling that scientific thinking is insufficient for addressing the being of our fellow animals and of humans. Need I add?: Philosophers, too, have typically fallen short in understanding animal being. An immediate disclaimer: Despite having been acculturated within a conservative Christian community as a youth, I do not pursue a Christian understanding of being, or any form of theism that includes an unquestioned ground, although the religion-based pacifistic teachings that I absorbed as a youth continue to influence me.

Where might a biologist like myself get guidance for such a task? For the last dozen years before retirement as a professor at the University of Maryland, where I guided a laboratory in evolutionary biology (specifically: molecular phylogenetics of arthropods—see Regier et al. 2010, 2013), I stumblingly tried to focus my thoughts by offering small discussion-based classes to undergraduates in the University's Honors Program. The topics of these classes were: the relationship between the theory of biological evolution and Christian theology, the philosophy of evolutionary biology, ideas about human selfhood, and animal ethics. Then, in retirement (2013–) I started reading more broadly in philosophy. Eventually, I came upon the writings of Martin Heidegger, and a book project started to come into my view, much faster—I might add—than my understanding of Heidegger, which remains a work in progress. So, like Christine Korsgaard, from whom I have taken inspiration, I am building my own ideas on top of those of an earlier philosopher, although one whose interests are more attuned to being as it relates to the human than to our fellow animals. (To be clear, my interest in being [see preceding paragraph] did not precede my own discovery of Heidegger; rather, it was his major gift to me.)

Heidegger considers his thinking about being to be *proto*ethical, indicating (at least to him) that it is conceptually prior to ethical thinking. This makes some sense because—as Heidegger says—being is not a being, while ethical theories are foremost about entities.[2] While some describe Heidegger as an ontologist, Heidegger himself describes his thinking—at least over part of his career—as hermeneutic phenomenology (i.e., as providing an interpretative stance toward what presents itself as itself, namely, in its very being). In his best-known book, *Being and Time* (Heidegger 1962: 25–6; originally published in 1927), Heidegger says about being: "Being—that which determines entities as entities, that on the basis of which entities are already understood, however we may discuss them in detail." I devote much of my book to thinking about being, particularly in the context of our fellow animals, but I note upfront that Heidegger rejects idealism (since being is neither an idea nor an ideal) and object naturalism (since being cannot be thought of as an object, but see Price 2013 on subject naturalism), and he considers theism to be outside his purview, especially when a god is thought of as an ultimately grounding being. It is also important to note that for all his talk about being, Heidegger does not deny embodiment—that would have been a non-starter for my interest in him—but he argues that it should be seen as derivative to a more fundamental understanding of being and way of being.

Being is fundamental to understanding humans and our fellow animals, but it should not be equated with the term "life," particularly when life is thought naturalistically as a material structure (a "body") or its process (a "mechanism"). Even so, being is always the being of some*thing*; that is, being is in beings but is not a separable *part*. Heidegger argues that we encounter bodies and their activities only through a prior understanding of being, even though there is no direct sensory access to being itself. Heidegger's thinking about being complexifies the traditional premise of biologists that they can provide an outsider's unimpeded perspective on the natural world. Within a Heideggerian frame, it is fair to say that before we can understand what embodiment and subjectivity might mean when applied to animals, including humans, we must understand their being. Understanding being is the task that Heidegger sets for himself, initially human being, then animal being, and later being more generally, although still in necessary relation to the one creature who he thinks can uniquely know being—the human.

In turning to Heidegger and to being, have I diverted my gaze from the ongoing devastation of our fellow animals? What is the connection between an understanding of their being and environmental devastation? Is this devastation fundamentally an ontological problem; that is, is it a manifestation of being itself? And here I want to be clear: the idea that a "correct" understanding of being must lead to an acknowledgment of environmental devastation—let alone a remedy for it—is misguided. To think in terms of necessity ("must lead to …") and causes is to frame the devastation in strictly scientific terms. Also, "correctness" reflects a particular—basically scientific—way of thinking about beings and does not apply to being. Instead, Heidegger talks about authenticity and about sheltering the truth of being. But though scientists do not talk about the truth of being as such, they still are among the most adamant in their concern for the ongoing environmental devastation. Does this suggest that being need not have any relationship to the devastation? Here it is worth pointing out that while many scientists are concerned about catastrophic environmental degradation leading to severe consequences for human civilization and to loss of biological species, their concern for our fellow animals as the individuals they are is seldom voiced; for example, consider the fates of individual research animals. Why is their moral concern for our fellow animals applied to species and habitats but seldom to individuals, while their concern for humans is squarely with individuals? Does this disparity relate to their understanding of animal being and human being?

Heidegger's project is not to understand the environmental devastation but how the meaning of being—and, closely related, of world—arises. Heidegger is not asking standard scientific questions about beings; instead, he describes his endeavor as the phenomenology of being, which prioritizes interpretation over description and prescription. According to Heidegger, there is a devastation in being itself that precedes and is more fundamental than its manifestation as the phenomenon of environmental devastation (Rentmeester 2016). The primordial devastation arises from our own naturalistic understanding of human being. Many—myself included—find Heidegger's thinking on human being compelling, but I have come to believe that he misinterprets animal being and its relation to human being. In so doing, he—and the many who share (mostly unknowingly) his misinterpretation—contribute to the ongoing devastation of our fellow animals. But some of the conceptual framework that he articulated for human being remains fruitful—I think—for rethinking animal being. Explaining this is the central task of this book.

So what is it about scientific thinking and naturalism that Heidegger finds objectionable? From a naturalistic perspective, we humans are presented with transformations in objects in a world, but Heidegger thinks, what it means to be in such a world remains outside our questioning. Instead, we might engage in political maneuverings to alter the arrangements of these objects (e.g., "solution-based thinking") while being guided by comfortable, preexisting categories that are left unquestioned. The question of the meaning of being (later, the truth of being) is Heidegger's question, and the arousal of this question is what motivates his writings. If this question is dismissed as meaningless from the outset, then there is little reason to engage with Heidegger's thinking. Nevertheless, I have adopted this question to see where it leads, even as I take a different look at our fellow animals from that of Heidegger. While there is much with which I disagree in Heidegger's thinking, he—among all the philosophers I have read—most invites in me a questioning mode to rethink the animal problem. Strongly supporting my endeavor, the written records of animal ethology and evolutionary biology have expanded dramatically since Heidegger's time, and although I do not claim that they can ground a new understanding of being—such would be counter to the way of Heidegger's thinking that I am trying to emulate—I believe that they invite an important path to rethink an animal's way of being. It would also be counter to Heidegger's way of thinking to suggest that a new ethical theory might result. Rather, a more humble aim is to develop a new understanding of an animal's way of being as a prelude to thinking about the animal problem that today rightfully motivates ethicists and environmentalists.

Heidegger's main treatise on the being of our fellow animals is *The Fundamental Concepts of Metaphysics: World, Finitude, Solitude* (Heidegger 1995; based on lectures from 1929/1930; abbreviated as *FCM*). In *FCM*, Heidegger devotes a large portion to the consideration of our fellow animals within the context of his central interest in world as experienced in human *Dasein*'s way of being (N.B.: For now, we can say that "human *Dasein*" is early-Heidegger's way of thinking about the human being as sense-maker in a radically different manner from that presented by scientists. Similarly unusual is his understanding of "animals" and "world.") Elsewhere in his extensive oeuvre, Heidegger mentions our fellow animals here and there, oftentimes with much repetition and never systematically. But Heidegger's ontological thinking has uniquely stimulated me to think differently than I previously did. His focus on "world" and "way of being"—rather than on, for example, values, worldview (perspective), or sentience and sapience (and other largely psychological terms that isolate the individual)—directs his thinking toward human *being* itself and how human existence can be either authentic or inauthentic. Unfortunately, his understanding of our fellow animals is much thinner and more problematic.

Alas, Heidegger's philosophy is not straightforwardly accessible to the general reader, including myself when I started this project, and this difficulty has been compounded for me by reading him in translation. Heidegger uses numerous neologisms and assigns radically different meanings to some words in normal usage. I am grateful for several excellent Heidegger lexicons (Dahlstrom 2013; Schalow 2020; Wrathall 2021). Heidegger's philosophy is particularly influenced by his understanding of the philosophers of Ancient Greece and Kant, who, according to Heidegger, almost immediately get sidetracked from philosophy's central issue of being. Even so, these other philosophers remain centrally important as Heidegger's own thinking emerges out of his understanding of that of his predecessors. I acknowledge upfront that giving adequate credit to these other philosophers is beyond my own ability and the scope of this book.

So, let's begin: Martin Heidegger the philosopher seeks (1) to rediscover being as presented originally in Ancient Greece, (2) to understand how being became transformed over the ensuing millennia, and then with all this (3) to think futurally—and possibly, to prophesy—from the inception to new and authentic possibilities of being. But where to start? In *Being and Time*, Heidegger suggests that an exploration of human *Dasein* is an appropriate starting point because—he claims—only human *Dasein* can think being and because *Dasein*'s own being reflects being more generally. But Heidegger strongly resists the existentialist (and naturalistic) interpretation of *Dasein* that some of his readers attributed to this early work, and so his thinking reorients in the ensuing years to other ways of thinking and saying being, although still necessarily linked to *Dasein*. One early way was to contrast *Dasein*'s way of being with that of our fellow animals, a theme he explores most extensively in *FCM*.

But a problem remains for anyone who tries to think being along with Heidegger: How can we think being when being is not a being? What vocabulary should we use when references to beings permeate our language so thoroughly? From *Being and Time* onward, Heidegger strives to develop an appropriate vocabulary, but still, as a teacher, he also wishes to communicate clearly to his students—to bring them along, starting from their familiarity with beings. One way he chose to do this was to shift the talk from beings as objects (humans, animals, etc.) and their underlying mechanisms to their *way* of being, thereby providing a link between beings (i.e., through their "way") and being itself without subverting the ontological difference. Briefly explained—at least as I present it—"way of being" can invite an ontological *interpretation* of naturalistic observations of animal behavior and evolutionary changes. Heidegger says that being is "that on the basis of which entities are already understood." Thus, understanding being—and way of being—is part of what it means to be, and this implies that there must (also) be someone who understands, that is, a knower. But a knower is an entity, albeit a special sort of entity who can know being. A central question in my book is whether our fellow animals can be such knowers, or whether such knowing is confined to humans, as Heidegger claims in *FCM* and elsewhere. Of course, we must address what it means to know. So, how can I accomplish all this?

In Chapters 1 and 2 of my book, I give voice to Heidegger's thinking about our fellow animals over his entire career, while still including my own critical commentary. In Chapter 3, I consider what Heidegger fails to grasp when he disclaims evolutionary thinking as relevant to making ontological claims about animals. I conclude Chapter 3 by presenting my own six ontological theses that compare and contrast the inanimate, our fellow animal, and the human in contradistinction to the three theses presented by Heidegger in *FCM*. In Chapter 4, I use Heidegger's own ontological framework of *Dasein*—namely, temporality and spatiality—to reinterpret the ethological record of a small group of bees that Heidegger interpreted in *FCM*. Then, I turn to the many ethological studies of these bees that post-date *FCM* to strengthen my claim that they reside in *Dasein*. In Chapter 5, I expand my ethological interpretations to numerous other animals (i.e., other arthropods, cephalopods, fish, birds, and mammals), albeit less intensively than for bees, but still making the case that they too reside in *Dasein*.

I am led to claim that Heidegger's understanding of our fellow animal's way of being is a misinterpretation of the ontical record.[3] That way he calls "captivation," which refers to an animal's absorption in itself and a drivenness and, correlatively, an animal's possibility for behavior only within an environment and not within a world. Instead, I argue that Heidegger should have meaningfully connected an animal's way of being with *Dasein*'s way of being, which he calls "existence" and which opens up the possibility for comporting within a world. For Heidegger, comportment implies a knowing-awareness in a world and concern for that world rather than a mere behaving in an environment. In making my claims about captivation and existence, I argue that Heidegger has four misunderstandings: (1) that all nonhuman animal behavior is instinctual,[4] (2) that instinct which, as a scientific (ontical) concept, refers to beings, is necessarily linked to ontological captivation, (3) that the way of being of our fellow animals is not *also* existence, and (4) that the human way of being is existence only, and not also captivation. My main claim, then, is that—at any particular time in history—the way of being of numerous species of animals, including the human, is plural, that is, existence in *Dasein* but also captivation. In saying this, I remain neutral as to whether being itself is plural.

To my way of thinking, all individual animals remain candidates for comporting and existing in a world. Why do Heidegger and I differ? I argue that Heidegger was hindered by a depauperate understanding of the available ethological record *and* a lack of openness to the ontological implications of ethological studies when thought in an evolutionary context. Even so, it remains clear enough without the theoretical underpinning of evolutionary biology that there is a taxonomically broad array of animals whose way of being is existence—or so I try to justify—and that this extends beyond the vertebrates, although animals like Porifera (i.e., sponges) and Placozoa may seem unlikely given their lack of indicators of whole-body integration (e.g., a brain, an internal skeletal system, whole-body polarization). As per Heidegger, existence as a way of being is categorically different from animal agency or sentience, psychological terms that have been used by others to distinguish among living creatures, so I do not use these terms either.

With existence *and* captivation as ways of being for (many of?) our fellow animals, what about creatures other than animals? For example, what about multicellular plants and fungi? What about multicellular forms of eubacteria and archaea, for example, biofilms? How should we interpret the group behaviors of unicellular prokaryotes, for example, their quorum sensing? As Heidegger himself often noted, commonsense understandings should be treated skeptically. One such widespread presumption is that a complex nervous system is a prerequisite for comportment, and since nervous systems are found in animals only, no other creatures could even have the potential to exist in Heidegger's sense. I think this conclusion is premature. Strong ontological claims about creatures outside Animalia should await further scientific studies. (For selected accounts of the status of plant intelligence, see DeSalle and Bachor 2020; Bakker 2022; Calvo 2022.)

With existence now an ontological feature of our fellow animals—and not just of humans—Heidegger's understanding of the entity called "*Dasein*" must be altered. *Dasein* now extends to our fellow animals, but as *Dasein* is an entity, there are indeed multiple forms, for example, human *Dasein* and social-bee *Dasein*. These multiple forms result from *Dasein*'s different *ways* of being. Even so, *Dasein* remains unitary when considered ontologically; that is, when considering *Dasein*'s way of *being*. Each form of *Dasein* has its own key features (called "existentialia"; singular: "existentiale"), but across forms there must be overlapping existentialia. Existentialia articulate aspects of *Dasein*, but aspects are never parts. Some existentialia are more fundamental (e.g., disposedness, projection, presencing) than others (e.g., fallenness, having a conscience, making room as found in human *Dasein*), but existentialia are not to be confused with the properties or characters as understood by biologists; rather, they are units of meaning-making. Articulating these units for animals is a work in progress. Similarly, *Dasein*'s different forms are not to be confused with biological species, or any other higher-level taxonomic grouping. Identifying forms of *Dasein* is an incomplete and ongoing project that requires much more attention to ethology (including evolutionary ethology) than is possible in this one book.

Once animals (including humans) are understood as living in existence and in captivation as their ways of being, a new world can open up for us humans, one that is impregnated with increased meaning, not in a psychological sense but in the sense that the being of our fellow animals—and not simply that of ourselves and other humans—now essentially matters to us. Conceived ontologically, this mattering is not an ethical claim but a claim about us as having the possibility to engage a world. Further, we can come to recognize that something similar for our fellow animals has been there all along. Whether biologists, including evolutionary biologists, care to reorient their questioning and research along these lines and what that reorientation might look like are unclear, but the increasing prominence of ethological investigations with emphasis on an animal's individual comportment is a promising sign that this is occurring.

In reinterpreting Heidegger, I am not claiming that his ontological thinking is not correct in some scientific sense; rather, animal behavior can be interpreted and biological evolution can be understood differently than he did, and this has important implications for interpreting an animal's way of being. Would Heidegger himself have been persuaded by my interpretations of new ethological and evolutionary studies? I think not, but likewise, I do not think Heidegger "saw" our fellow animals and their way of being clearly enough.

Lastly—and returning to the matter of environmental devastation—Heidegger's philosophical views of our fellow animals have been interpreted as justifying the status quo as regards animal rights and welfare, and this has warrant. Even so, he only briefly and obliquely discusses such issues, plus he clearly argues against thinking hierarchically about animals (including humans). In his later writings, however, he claims that positionality/enframing—what he calls the being of modern human technology—has led to industrial-scale farming with devastating consequences for humans and (other) animals, although nowhere does he directly connect positionality to a nonhuman animal's way of being, and indeed it is not clear that there can be a connection given his understanding of their being. In other words, I interpret Heidegger as saying that "devastation" refers to (1) *human* beings and (2) human *being* and (3) *animal* beings, but not to (4) animal *being*. Here, then, we see a hint of a negative constraint of Heidegger's understanding of being on the world we inhabit—a constraint with ethical consequences.

How can one evaluate an ontological claim when one is talking about "no-thing" (i.e., being), and is evaluation even the point? Do we gain something by addressing being explicitly, or is it as meaningless an exercise as the positivists of Heidegger's time claimed? I believe there is something to be gained—something essential for us as the individual humans we are in a world shared with our fellow animals. To "see" the truth of being is challenging because being is no "thing" for which we have developed a comfortable discourse, and so we can easily be led astray. But how else can we reason through our moral and ethical claims without already having some sense as to how the world is? My task in this book is to think ways of being of many different animals. My direct engagement with the "ought" of morality and ethics is minimal. That is a path for another book.

Heidegger's thinking has provided me with numerous robust challenges in ways that other philosophers have not. That is why I was first fascinated by and remain grateful for his thinking. I invite you—the reader—to join Heidegger and me in thinking about our fellow animals.

ACKNOWLEDGMENTS

Janet Friesen Regier—my wife of fifty-six years—has been my constant companion and biggest booster while writing this book. She is the only person who read and commented on the entire manuscript prior to submission to the publisher. To her, I lovingly dedicate this book. I thank our two children—Emily and Jeff Regier—for their long-term support and encouragement, and Emily, for her helpful comments. I thank Pumpkin and Mineral for showing me who they were and for providing me with so much joy.

My first exposure to Martin Heidegger occurred as an undergraduate at Harvard, when I observed my roommate (Magne Warren Kristiansen) reading *Being and Time*. Over the ensuing decades, I have forgotten neither Magne nor that first glancing exposure to Heidegger. However, my more immediate interest at that time was to become an experimental biologist, and for this, I thank Professor Fotis C. Kafatos, my cherished mentor while I was an undergraduate, graduate student, and postdoctoral fellow at Harvard. Fotis taught me how to think as a scientist and to appreciate Lepidoptera, both of which made this book possible. Even back then, I did have other academic interests, particularly at the interface of science and religion, and for fostering those interests—which served as a gateway to my eventual engagement with Heidegger—I thank my long-time friend and informal advisor Gordon D. Kaufman—formerly a professor of philosophical theology at Harvard Divinity School. Eventually, I too became a professor, and while my research was strictly biological, my teaching was broadly directed to science and religion and to animal ethics, thanks in large part to encouragement from Professor Charles Mitter. For our long-term friendship and numerous rewarding research collaborations, and for his many professional encouragements, I remain deeply grateful. I acknowledge the Department of Entomology (University of Maryland) and the Center for Biosystems Research (University of Maryland Biotechnology Institute) for providing me space and freedom to do research and teaching, and for colleagues who greatly enabled my exploration of ideas during my twenty-five years at Maryland. Upon retirement in 2013, my wife and I returned to Cambridge, where the then-Dean of Harvard Divinity School—Professor David N. Hempton—graciously offered me a one-year appointment as Visiting Scholar. This allowed me to reintegrate into the intellectual life of Harvard, thereby making my writing project much more fruitful. At Harvard, I have been particularly influenced by the writing and teaching of Professor Christine Korsgaard—from whom I absorbed the idea of using the writings of a specific philosopher as a platform from which to explore my own thoughts about animals—and Professor Sean Dorrance Kelly—from whom I absorbed an enthusiasm for Heidegger's thinking. I thank members of the Dialectical Thinking group within the Mahindra Humanities Center at Harvard for valuable discussions. Finally, I thank two anonymous reviewers of this manuscript for their helpful comments.

ABBREVIATIONS

Being and Time	*Being and Time*: Heidegger 1962
Contributions	*Contributions to Philosophy (Of the Event)*: Heidegger 2012
FCM	*The Fundamental Concepts of Metaphysics: World, Finitude, Solitude*: Heidegger 1995

Introduction

In this book, I aim to seriously engage Heidegger's thinking, and especially his thinking about animals; hence, my book's principal title: *Heidegger and Animals*. I have an overarching reason for doing this—I wish to push against his claim that there is an ontological abyss separating our fellow animals and humans, and I do this principally by (re)considering scientific discoveries from ethology and evolutionary biology; hence, my book's subtitle: *New Insights from Ethology and Evolutionary Biology*. The arguments I garner against his abyssal thinking are not, and cannot be, solely based on biological thinking, since Heidegger's claim concerns the being (or "way of being") of animals, and not animals as beings or, said scientifically, as organisms. So, when I say I am offering "new insights," what I mean is that I intend to follow in Heidegger's tracks by offering a new *interpretation* of an animal's way of being from the perspective of my own scientific understanding of ethology and evolutionary biology.

Unlike many philosophers and people generally, Heidegger (b.1889–d.1976) directs his thinking toward being rather than beings. This he calls "ontological thinking" (i.e., thinking being) as opposed to "ontical thinking" (i.e., thinking about beings). Thinking about or understanding beings is what scientists do. Thinking being, he says, is what philosophers (ought to) do. While Heidegger's understanding of being evolves, a particularly crisp definition is given in *Being and Time* (Heidegger 1962: 25–6): "Being—that which determines entities as entities, that on the basis of which entities are already understood, however we may discuss them in detail." For those who insist on interpreting his writings as fundamentally about entities (e.g., societal norms) and their presentist relationships, or about human consciousness, as is common in existentialist interpretations of *Being and Time*, his overall thinking thereby becomes either misinterpreted or inscrutable. The difficulty Heidegger confronts in making his thinking understandable to readers is reflected in the paucity of words that refer to being. As a result, Heidegger improvises, both by assigning new meanings to old words and by using innovative punctuation within words—most commonly dashes—to indicate new meanings. To understand Heidegger, it is important to know what he means when he uses such words. Many of these words are important to my own project and, so, a discussion of some key words will constitute this Introduction.

Let us begin with "animal" and its biological understanding. I start this way because, as an evolutionary biologist—but now also a neophyte philosopher—I am interested in how and why the thinking of biological researchers differs from that of Heidegger. Are they non-overlapping, and if not, how do they relate? For biologists, "animal" identifies a subset of living beings who are embodied, potentially metabolically active, and self-sustaining beings capable of imperfectly reproducing themselves and evolving—broadly speaking—in a Darwinian fashion. One animal species is demarcated from another by having a diagnostic suite of characters, where "characters" refer to specific substructures, processes, or behaviors. Living beings are distinguishable from nonliving entities, as only the former evolve in a Darwinian fashion (see Wong et al. 2023), even though both are of the same material order. But if this understanding of animal is grounding for biologists even when they are not engaged in research, that is, when they are more broadly considered as members of society who think about, teach about, and interact with animals, then, for them, animals simply are active entities who evolve in a particular manner. If this is the case for you, and if you have no interest in thinking about that strong metaphysical claim, then Heidegger's lifelong project of thinking being will be of no interest, and you can stop reading this book. But it seems obvious to me that many people (myself included) live their lives in a manner that occasionally contests the strongly materialist interpretation that I have outlined above as the sole understanding, and I am interested in thinking more deeply about such contestations. I do not do this by challenging materialistic thinking itself. Instead, I pursue a line of inquiry that takes seriously my own thinking about biological research and Heidegger's writings as they relate to an understanding of an animal's way of being.

But first, are all approaches that we call "biological research" equally capable of informing us about animal being? I think not. For purposes of this book, let me outline three broadly overlapping approaches present in modern-day biology: (1) Within *evolutionary biology*, systematic biologists identify evolutionary groupings (e.g., of genes, populations, species, genera, etc.) based on inferring shared derived characters[1] in order to build evolutionary trees of life. As such, these biologists are inherently interested in the diversity of life forms. Such trees are metaphorical representations used to compactly describe genealogical relationships among multiple taxonomic groupings. Other evolutionary biologists document the structural, functional, and behavioral consequences of character evolution that takes place within populations of individual animals. In this book, I present and interpret results from the evolutionary subfields of evolutionary ethology and ethological paleontology, but I do so with a sensitivity to Heidegger's thinking (see Chapter 3). In evolutionary biology generally, time—considered over many lifetimes—and place, or environment, are central to understanding character evolution. I highlight the roles of time and place (space) in evolutionary research because both are important concepts for Heidegger, although his understanding of them is different. (2) The various subfields of *structural and functional biology* (e.g., biochemistry, morphology and structural biology, physiology, cell biology, developmental biology) focus on the characters themselves as found within individual animals. Time remains relevant here—enzymes have kinetics, cells have life cycles, and organisms have life histories—but still, time is not seen as having the explanatory power that traditional evolutionists assign to it. Instead, there is more talk about "mechanism." I will have little to say about this research approach (but see Afterword). (3) *Ecology* and *ethology* occupy a temporally intermediate position relative to #1 and #2 above. Ecologists typically study organismic interactions and adaptations within local communities, and are not as reductionist as structural and functional biologists, who focus on the suborganismic, but neither do they focus on higher-level groupings nor very-long-term changes as evolutionists do (exception: evolutionary ecology, see also the long-term ecological results from the National Ecological Observatory Network [https://neonscience.org]). In ethology, which could be considered a subfield of ecology, researchers typically study the behavior of individual organisms. When behavioral studies take place in natural environments, the boundary with ecological studies blurs. Ethological research can also be carried out in more controlled laboratory settings (experimental ethology) and over multiple generations (evolutionary ethology). Interpretation of ethological research features prominently in this book (see Chapters 4 and 5).

Heidegger rejected both evolutionary and suborganismic studies as ontologically misguided with respect to understanding animal being, but for now I want to point out that he was sympathetic to ecology and ethology, although only to a degree. For Heidegger, the interesting point about ecology and ethology is their focus—particularly for ethology—on the individual being as it relates to its environment, neither deconstructing the individual into parts nor burying it in higher-level categories. Over his long career, Heidegger directs his own thinking toward the individual human but also toward "a people." We might speculate that the individual nonhuman organism and its interbreeding population, respectively, provided evocative homologs of the human situation for Heidegger, but this would be wrong. Instead, while ethologists and Heidegger are both interested in how individuals and their meaningful groups exist in their environment, this similarity is only superficial because "individuals," "meaningful," "a people," and even "environment" for the later Heidegger all have different meanings from that used by ethologists, so we come up against a semantic challenge in inter-relating the two thought-systems. Even so, we can generally say that more reductionist biological accountings are farther removed from Heidegger's understanding of "animal."

An early ecologist/ethologist who influenced Heidegger was Jakob von Uexküll (1864–1944), who incorporated both materiality and relationality into his understanding of the animal. For numerous animal species, von Uexküll shows how an animal can form a particular web of relations with what is outside itself, thereby constituting that animal's world (von Uexküll 2010). Of interest is that von Uexküll does this while largely rejecting Darwinian evolution and its principal mechanism, natural selection.

Although somewhat sympathetic to von Uexküll's understanding of the animal in its relationality, Heidegger from *Being and Time* onward almost entirely avoids talk of materiality (or embodiment) and instead talks about "animal" more formally, for example, in *Fundamental Concepts of Metaphysics* (Heidegger 1995; abbreviated as *FCM*) as a locus of "manifold instinctual activities" (p. 248). In doing this, he is not denying animal embodiment, only that it is not fundamental to understanding the animal and its being. Even as regards relationality, Heidegger rejects von Uexküll's more physics-centric understanding of this as interactions between discrete objects ("organisms"), particularly when applied to the human animal, as von Uexküll does. To Heidegger, human relations are more than simply presentist interactions. They point in a projective sense to human *being*, something that von Uexküll not only ignores but implicitly misunderstands, or so Heidegger claims. As concerns relations among our fellow animals, Heidegger is more circumspect. These animals are neither humans, who have relations, nor inanimate objects, who have only interactions. But what are they?

Von Uexküll presents Heidegger with two more equally fundamental problems: (1) von Uexküll considers both humans and our fellow animals to have worlds, albeit distinct, and (2) von Uexküll considers different species of animals to have different worlds. By contrast, Heidegger claims (1') that only humans can have a world, and (2') that all animals have the same being and all are "poor in world," or even without world. In light of this, we'll want to consider in depth Heidegger's distinct understanding of "world," why is it that our fellow animals are excluded from such, and how "world" differs from "being."

In Chapters 1 and 2, I'll have more to say about Heidegger's own understanding of "animal," but in the spirit of Heidegger, I'll end this discussion of "animal" with a metaphor that captures that spirit. Consider all the interactions between our fellow animals as branches of a single huge dynamic network, but think of them—at least according to my interpretation—not as actualities but as possibilities that are unstable over time and space. Then, consider the points where they intersect as the transitory and potential significances of these possible interactions. These significances "are" the animals themselves, but never to be thought of in isolation from the network. Where the intersecting branches are antagonistic, for example, between predatory behavior and avoidance behavior, we can talk—but only in a derivative sense—of there being embodied beings, for example, a spider as predator and a fly as prey. Where the intersecting branches are mutually supportive, for example, between courting behavior and positive evaluative (receptive) behavior, we can talk about there being other kinds of embodied beings, for example, two mating swans. The understanding of "animal" comes from appreciating the entire network, not just the nodes, but not as something material. In my view, Heidegger's trans-temporal and trans-spatial thinking about animals can be considered open to ecological, ethological, and evolutionary thinking, although he brings a necessary holistic focus that is in tension with object-based thinking in biology. So much for my preliminary consideration of "animal."

"An" may seem a strange word to discuss in a book's introduction, but for Heidegger it is loaded with portent. Heidegger talks about *the* animal's way of being without distinguishing it from *an* animal's way of being, and this is because he claims that all animals (excluding humans, for whom—he thinks—it is fundamentally misguided to refer to as animals) have the same way of being—

what he calls "captivation." In fact, when he discusses various animal species in his major treatise on animals (*FCM*), Heidegger uses the scientific grouping "animal" more broadly than our current understanding; for example, infusoria (unicellular protists) are included. By contrast, when referring to the human way of being—what Heidegger calls "existence"—"an" is not just allowed but required. This is because humans are selves, and selves are individuals, whether—to use Heidegger's language—they are inauthentic (e.g., accepting someone else's way as their own) or authentic. So, existence comes in "flavors" and is connected to an individual. Even when the ways of being for two humans are seemingly identical to an outside observer (say, to a scientist), Heidegger says that each human has a way of being that is of special significance to himself/herself. By comparison, an animal's way of being has significance only in relation to and for humans, and that significance relates identically to all individuals within its species.

At this point, I need to blunt potential misunderstandings. Heidegger does not deny that biologists can meaningfully say, "*an* organism," but here "organism" is a biologically defined entity and "an" refers to a particular number, namely, one. Heidegger's understanding of "organism" is different (*FCM*: 235), and "an" and "individual self" are not terms of simple numericity but ones that function as units of significance.

Now, let us consider "way," "being," and "way of being." The latter is one of several terms Heidegger uses to refer to the being of entities. As a prelude, we can say that Heidegger does not directly address who the animal or human *is* in her constancy, universality, and/or substantiality as a biologist might. Instead, Heidegger is interested in *how* it is to be a human or an animal; that is, what a human or animal must *do* in order to be called a human or an animal. Prioritizing doing indicates why ethology, with its emphasis on behavior, is of particular interest to Heidegger. "Way" is a metaphorical term that relates to this doing and illustrates both a path that is followed (the way followed) and an "activity" of following that path (being's way of following a path); that is, way has both static and dynamic senses, but these are not to be confused with physical things (e.g., a drive*way*) or with actual physical movement. Heidegger also attributes dynamic and static senses to being itself, when he describes being through the Ancient Greek understanding of *phusis* as the *emerging-abiding* sway (Heidegger 2014). So it is appropriate that Heidegger combines "way" and "being" into a single term—"way of being." Importantly, "way of being" is a phrase that, in a limited sense, remains open to the inspection of entities (e.g., by observing the comportment of humans, the behavior of animals, and the properties of stones). But "inspection" is not only a visual claim about a being but also requires what Heidegger calls an "ontological interpretation," that is, an interpretation of being itself. So in practice, the challenge of understanding an entity's being and its way of being is not distinct. In due course, we will consider what an ontological interpretation might be, but for now we can say that "way of being" points to being *and* beings, and that humans, animals, and stones (to use Heidegger's partitioning in *FCM*) have distinct ways of being. Considering ways of being is useful for making comparisons and contrasts of entities. Heidegger talks about different ways of being when discussing certain large-scale human comportments, for example, the way of being a biologist or the way of being a physicist (Heidegger 1962: 33). He also talks about the animal's and the stone's distinct ways of being, but always in relation to the human as a thinker of those ways (*FCM*: 198, 231).

Heidegger sometimes uses a different term—*kinds* of being—when talking about being more generally, for example, the being of humans, of (other) animals, and of equipment (*Being and Time*: 32, 98). "Kinds" evokes a sense that differences in the being of these beings (human, animal, equipment) are fixed and unbridgeable, unlike the "ways" expressed in different human comportments, although sometimes he uses both "kind" and "way" when referring to stones (*FCM*: 204). Heidegger also uses "manner of being" (*FCM*: 197), "mode of being" (*Being and Time*: 26–8), "region of being" (*Being and Time*: 145; *FCM*: 354), and sometimes he refers directly to an entity's being, for example, "animal being" or "human being." Absent clear or consistent guidance, I avoid parsing these different ways of referencing being and, instead, stick with "way of being" when referring to animals, including humans.

In talking about way of being, it is important to emphasize that it is Heidegger himself who is doing the thinking, so "way" refers both to a way of being of the animals, for example, as presented in *FCM*, and to Heidegger's way of thinking an animal's way of being. Heidegger is clear that both ways must be present in their unity for there even to be being, with the further understanding that thinking is frequently not explicit, although it can be as, for example, when Heidegger was writing *FCM*. In terms of thinking being ("essential thinking"), Heidegger contrasts his way with representational thinking that focuses on "proper" procedure (e.g., the scientific method) and already anticipated outcomes (e.g., scientific hypotheses that are based on prior representations). For Heidegger, essential thinking entails that "there are no paths laid out in advance" (Heidegger 2015: 145). The essential point to understand here is that it is questioning itself—along the way—rather than any particular outcome or conclusion that constitutes a real way of thinking. But now something similar can be said about the way of being. Our understanding of it must remain open to interpretation, and not falsely rely on a search for deductive proof. Yes, experimental observations are important, but essential thinking also requires "seeing" outside of beings themselves (e.g., the organisms) and toward their being, with further interpretation always a possibility.

Now let us briefly consider "being." To repeat: Heidegger says, "Being—that which determines entities as entities, that on the basis of which *[woraufhin]* entities are already understood." Here, "determines" should not be thought causally, since causation applies only to beings. But also, it is clear that understanding is central to a being's being. Understanding is the way by which being can appear, and that way is to be distinguished from a direct encounter ("causation") with an entity; rather, understanding being requires interpreting entities. Absent understanding, there are only entities, that is, no being. Later, I push back against Heidegger's claim that an animal's being only makes sense when there is a *human* interpreter. Instead, I claim that our fellow animals can themselves be interpreters of their own being as well as that of other animals, including humans with whom they are sometimes associated.

Heidegger eventually finds the above "definition" of being to be inadequate, in part, because being is too tightly connected with beings. Indeed, he finds that adequate words simply do not exist to convey what being might be, although he never gives up trying. He devotes his entire career to understanding being, initially within a systematic philosophical framework (e.g., *Being and Time*, *FCM*) and later in more open forms (e.g., in his own thoughtful commentary on the poems of a few "enlightened" poets [e.g., "Language in the Poem": Heidegger 1971a] and in written "conversations" among friends [e.g., "*Agchibasie*: A triadic conversation on a country path …"; Heidegger 2016a]). In doing this, Heidegger is searching for new ways of thinking being (not to be confused with thinking *about* being) without ever claiming direct access to being itself.

Heidegger never gives up on the idea that there might be an underlying unity to being itself, even though this universalist thought would seem to conflict with some of his other claims, for example, that being maps to (human) selves and not to humans in general. For purposes of writing this book, pursuing that important thought about Heidegger's own thinking would keep me from my task of understanding an animal's way of being. My principal aim in writing this book is not to elucidate Heidegger's thinking of being itself but to reflect on Heidegger's understanding(s) of being—and way of being—with the hope that this can provide me with a path to my own understanding of an animal's way of being.

Let us briefly reflect on "thinking," which is a central term of art with numerous distinct, but overlapping, understandings (e.g., metaphysical thinking, being-historical thinking, inceptual thinking, commemorative thinking, mindfulness, knowing-awareness, enowned thinking, innovative thinking, calculative thinking, representational thinking; see Schalow 2020). Most centrally, Heidegger strives to make his own thinking post-metaphysical and, therefore—according to him—as moving beyond *all* philosophers since Plato, who in one way or another posited absolute grounds. The idea is that Heidegger's thoughts do not simply extend or modify a preexisting metaphysical framework but rather that they should no longer be considered metaphysical as this term is traditionally understood. Even so, Heidegger is adamant that he is not rejecting what came before. Instead, he sees himself as starting once again from the origin of Western philosophy—particularly as found in the pre-Socratics—and, later in his career, as thinking in a new way—what he called "being-historical thinking."

By now it should be obvious that since the aim of this book is to think about an animal's way of being by reflecting, in part, on Heidegger's own understanding, I will not be able to impose my own understandings of "animal," "human being," etc., which—until I started this project—could have been characterized as mainstream scientific. Rather, I must begin by understanding what Heidegger himself thought. This will not be an uncritical undertaking, but if Heidegger is to be meaningful for me to reflect on, then there can be no meaningful criticism of him without first understanding him.

Finally, I want to reflect a bit on Heidegger himself. Why did I choose Heidegger over any number of other philosophers, or non-philosophers for that matter? Of course, the choice is personal but still requires clarification. But before the philosophers, there are the animals, and animals are the central players in this book because, as one who is living in the midst of an ongoing extinction of life forms, I am disconcerted. Like Heidegger, who sees both the promise and the pitfalls of technological thinking, I am increasingly skeptical that our fellow animals will be able to thrive in a human-dominated world. But when I say "animal" and "thrive," I could still be speaking as a biologist—about beings. For Heidegger, technology is first of all an assault on human *being* itself. It is not just Heidegger's writings on technology that resonate with me as I think about our fellow animals; it is his focus on being itself that has opened up a new way of thinking for me, one that I find worthy of exploring. His challenge to technological (scientific) thinking and his rejection of traditional metaphysical foundations, including religious ones, warrant my consideration.

Later in his life, Heidegger develops the ontological concept of "letting-be" or "releasement" (*Gelassenheit*), which entailed a less "activist" account of human *Dasein*'s relation to being. Some have suggested that letting-be was his response to his own disappointment and to others' opprobrium for the way in which he "translated" his understanding of the human's way of being into a political mode of governance at the University of Freiburg, where he was Rector for a brief period during Nazi times. More fundamentally, letting-be reflects Heidegger's eventual understanding that being cannot be willfully changed. Rather, the human (1) must remain open to the possibility—and this is therefore always an uncertainty—that a more favorable situation might arise, then (2) to be able to recognize this unique situation if it arises, and finally (3) to allow being itself "to be" the guide for ushering in a new understanding of being. Heidegger, as the unabashed elitist that he was, thinks that few people [he calls them "the future ones" in *Contributions to Philosophy (Of the Event)*; Heidegger 2012b; abbreviated as *Contributions*] are able to accept the (ontological) loneliness that follows from rejecting the modern understanding of human *being* as (technological) enframement *(Ge-Stell)* and, instead, to remain open to being's possibility. Letting-be becomes his response to the ongoing technological assault on the human being and its way of being. It grounds a different sort of human response to technological crises than, say, technologists/scientists offer, which typically is simply more technological intervention. Heidegger saw technology itself as driving the crisis and saw humans increasingly as its pawns, or messengers.

Heidegger's letting-be is not detachment from the natural world. Rather, it is a serious attempt to let the world, as he understands it ontologically, reveal itself as it is to those few and future humans who have prepared themselves to be open to world change (see Mitchell 2015; Rentmeester 2016). Heidegger can be frustratingly obscure as to how a change in being might map onto real-world situations, but this is because a change in being can be neither deterministic nor fully understandable. Other than during the Rectorship period, Heidegger's writing largely avoids overt ethical claims (but see Habermas 1977; Schalow 2001; Kisiel 2001; Dahlstrom

2013: 67–9; Fried 2017). To think that there is a direct path from letting-be to some specific outcome concerning beings would be to misinterpret Heidegger, for example, as necessarily supporting modern environmentalism. Letting-be becomes another one of Heidegger's provocative ideas that offers me a new way of thinking about our fellow animals.

Central to my turn to Heidegger is the depressing observation that the many convincing ethical arguments which militate against the ongoing mistreatment of our fellow animals have oftentimes been met with disdain, distrust, or silence in the United States and other technology-driven countries. Naturally, these ethical arguments focus on animals as beings. I ask: What is so hard to understand about Jeremy Bentham's claim in the eighteenth century of animal suffering, Peter Singer's (see Singer 2002, 2023) modern rephrasing of that claim, or Christine Korsgaard's (2018) eloquent exposition that animals, like humans, have their own good, a good that ought to matter to us? Earlier in the twentieth century, American philosopher John Dewey (1926) provided a partial response:

> Scientific men are under definite obligation to experiment upon animals so far as that is the alternative to random and possibly harmful experimentation upon human beings, and so far as such experimentation is a means of saving human life and increasing human vigor and efficiency.

And,

> Surely, until it is finally decided that the taking of animal life for human food is wrong, there is something morally unsound in any agitation which questions the right to take animal life in the interests of life and health of man, women, and children, especially when infinitely more precautions are used to avoid suffering in the later case than in the former.

Dewey's lack of concern for our fellow animals, his obfuscation of the fact that animals too can flourish and suffer, is today even more apparent. Is it naive on my part to think that there remains an unresolved challenge here, namely: What we consider our fellow animals *to be* rather than the currently nearly exclusive focus on their utility to humans and how humans ought to relate to them? At least for me, Heidegger's thinking opens a previously unexplored path, one that is meant to serve as a *protoethical* foundation but not as an absolute ground. If this is the case, then understanding being—and way of being—becomes central to understanding our fellow animals as the animals they are. This book is an exploration of this unresolved challenge.

Finally, I note that my own thinking about our fellow animals, their way of being, and their world takes place within Heidegger's Western academic tradition of philosophical and scientific inquiry. It does not have to be this way. Other traditions present different ways of thinking. For example, in considering Anishinaabe and Potawatomi understandings of the world as lived in the modern-day United States, Kimmerer (2013, 2024) talks about respect, responsibility, reciprocity, and love. Hers is a compelling account, but it is one that I believe can also be found as a minor theme in my own Western tradition. Alas, exploring these connections is beyond the scope of this book.

This, then, might be an appropriate place to delimit the goal of my book. I have taken up one large question—whether or not there is an ontological abyss separating humans and our fellow animals. My approach to this question follows Heidegger's own thinking on human *Dasein* (briefly: that entity whose own being matters to itself in its knowing-awareness; a meaning- or sense-maker) and its way of being, that is, existence. Heidegger says there is an abyss because only humans can exist in *Dasein*. But in making my argument against his claim, I confront other large and closely related questions. For example, in claiming our fellow animals, too, can be in *Dasein*, I have not taken a position on whether that includes all animals, or only some, nor have I considered whether other forms of life (e.g., fungi) can exist in *Dasein*. Additionally, I have barely touched on captivation, which is Heidegger's term for animal being. The few claims that I have made about captivation require further explication and defense, particularly my claim that animals, including humans, exist in *Dasein and* live in captivation. Then, there is the question as to whether existence and captivation identify a fundamental ontological division, or whether an even more inclusive ontological understanding is warranted. This is where we might wonder how life/living relates to animal existence and captivation (e.g., see Thompson 2007; Barbaras 2021), but if I were to take up that topic, then I would also want to consider whether the living/nonliving dichotomy is fundamental, or not (Krauss 2012; Wong et al. 2023). I trust that I will not be judged too harshly for restricting this book's scope.

Here is a summary statement of this book. In Chapter 1, I discuss in detail Heidegger's most extensive and systematic thoughts about our fellow animals as presented in lectures of 1929–30, and later published as *The Fundamental Concepts of Metaphysics: World, Finitude, Solitude*. *FCM*'s overall structure, content, and approach are described, and this is followed by discussion of each of *FCM*'s three parts—PRELIMINARY APPRAISAL (on metaphysics), PART ONE (on boredom), and PART TWO (the "comparative" analysis of our fellow animals and of humans in their *Dasein*, or simply, human *Dasein*). All three parts together are framed by Heidegger so as to support his claim that there is an unbridgeable ontological abyss separating the human and human *Dasein*'s way of being (which he calls "existence") from that of our fellow animals (which he calls "captivation"). Even so, near the end of his comments on animals, he admits uncertainty about his abyssal thinking, but immediately thereafter, he simply dismisses this uncertainty, as is apparent from the anthropocentric focus of the culminating last chapter. But for the critical reader of *FCM*, the consequence of Heidegger's intellectual slippage is that grasping Heidegger's understanding of the way of being of our fellow animals remains in limbo somewhere between that of humans and stones (i.e., the inanimate), even though Heidegger does not acknowledge this slippage and even while his ongoing questioning remains essential to his way of understanding the human being.

In Chapter 2, I discuss numerous other works by Heidegger where our fellow animals are more briefly mentioned. (Heidegger started publishing in 1910 and remained active until his death in 1976.) I argue that (1) before publishing *Being and Time* in 1927, Heidegger has yet to articulate his own definite ontological understanding of our fellow animals, but (2) that by the time he delivers the *FCM* lectures two years later, his ontological understanding of our fellow animals—including the ontological abyss separating them from human *Dasein*—is fully formed and does not fundamentally change in the ensuing decades.

In the final three chapters, I—like Heidegger before me—present and interpret behavioral studies of specific organisms as ontologically informative, but—unlike Heidegger—I interpret these studies to indicate that our fellow animals, as well as humans, exist in *Dasein*. Specifically in Chapter 3, I relate Darwinian evolutionary thinking to *Dasein*, arguing that such thinking—particularly as guided through evolutionary ethology and ethological paleontology—provides a temporally expanded understanding of animal *Dasein* (now including human *Dasein*) as well as insight into previous changes in *Dasein*'s "forms" of existence. I consider such changes in *Dasein*'s forms to be an application of later Heidegger's historical *Dasein* concept. I argue that Heidegger could have provided a richer understanding of *Dasein* if he had been more receptive to evolutionary thinking. Chapter 3 concludes with my formal response to Heidegger's three comparative ontological theses in *FCM*, particularly the second one that claims "the animal is poor in world." The response—in the form of six theses—includes the proposal that humans (here to be distinguished from human *Dasein*) and our fellow animals are in both existence (i.e., *Dasein*'s way of being) and captivation. But, existence and captivation as the ways of being they are remain ontologically distinct because *Dasein*'s way of being—whether of the human or our fellow animal—is never captivation. Thus, Heidegger's claim of an unbridgeable ontological abyss remains, but it no longer separates humans and our fellow animals, who embody distinct forms of *Dasein* (and likely many more than two forms when considering the diversity of animals). The six theses together synthesize my specific interpretations of behavior as presented in Chapters 3–5.

In Chapter 4, I make determinate that some animals besides humans are in *Dasein* by interpreting ethological studies of a small group of extant organisms—namely, worker honey bees and bumble bees—whose behavior is considered by Heidegger to be solely driven by instincts. I critique Heidegger's interpretation of selected historical studies of bee behavior as corroborative solely of animal captivation. Then, I introduce numerous more recent ethological studies of bees and interpret them within the framework of *Dasein* and its way of being as existence. Specific bee behaviors are related to one or more of Heidegger's conceptually articulated "temporal moments" of *Dasein*'s existence—beenness, presencing, and projection—plus also spatiality.

In Chapter 5, I interpret the modern ethological record of numerous other animals as also in *Dasein*. Separate sections are devoted to arthropods in addition to bees, cephalopods, fish, birds, and (nonhuman) mammals. While arguing that Heidegger misunderstood an animal's way of being, I acknowledge that Heidegger's focus on (1) being rather than beings, (2) *Dasein* rather than consciousness or some other scientific understanding of the human, (3) existence as *Dasein*'s way of being rather than, say, sentience, agency, embodied experience or some other psychological way, and (4) captivation has gainfully contributed to my own way of thinking about our fellow animals, including their ongoing devastation by human technological thinking.

Reflecting the crisis of animal devastation is the lack of consensus in the West over what it means to be an animal. For example, consider the distinct understandings of "animal" in: Christianity (as a creation of a god), sub-organismic biology and psychology (as an internal physico-chemical or psychological mechanism), ecology (as an organismic interactor), and economics (as a commodity). Heidegger rejects all of these understandings as metaphysical, that is, as dogmatically grounded. Instead, in focusing on an animal's way of being, he is striving for a post-metaphysical understanding of being.

Unfortunately, I have been led to the view that despite all his seminal thinking, Heidegger unknowingly contributes to environmental devastation by unreasonably othering our fellow animals when he claims there to be an unfathomable ontological abyss separating our fellow animals and humans. Over his long career, Heidegger's own thinking evolves—from animals having an environing world (when early Heidegger was first reading and thinking about animal ecology and did not challenge the traditional understanding of "world"), to their having an (ontical) environment but no world (when Heidegger invoked the ontological difference), to their not even having an environment (as Heidegger moved away from the ontological difference and toward letting-be). I am not saying that his earliest version was right. This is because he had yet to explicitly consider an animal's way of being. What I want to claim is that had Heidegger been more attuned to the particular strengths of biology—particularly as found in ethology and evolutionary biology—and had he been able to interpret these disciplines with more understanding as to their connection to the ontological, he could have reoriented his abyssal thinking without eliminating it, and could have reasonably connected humans *and* our fellow animals to *Dasein*'s existence and to the captivation of all animals (including humans). I argue that his doing this would not have been a capitulation to a biological understanding, at least not when that understanding is neither highly reductive nor materialistic.

Chapter 1

HEIDEGGER'S UNDERSTANDING OF AN ANIMAL'S WAY OF BEING IN *THE FUNDAMENTAL CONCEPTS OF METAPHYSICS: WORLD, FINITUDE, SOLITUDE*

Section A. The General Structure, Content, and Approach of the Book

A quick reading of Heidegger's commentary on animal being in *FCM* can easily result in seeing its three parts—PRELIMINARY APPRAISAL, PART ONE, and PART TWO—as radically distinct: the first, on philosophical methodology, which is a long-term interest of Heidegger's; the second, on identifying boredom as a fundamental[1] attunement, supplementing his coverage of anxiety in *Being and Time*; and the third, on the being of human *Dasein* in comparison with that

To minimize possible confusion while reading my text, the classificatory schemata of *The Fundamental Concepts of Metaphysics* (Heidegger, 1995; first presented as lectures in 1929–1930) are presented for comparison.

Fundamental Concepts of Metaphysics	***Heidegger and Animals***
PRELIMINARY APPRAISAL	Chapter 1
Chapter One (§§1–3)	Sections A–D
Chapter Two (§§4–7)	Chapter 2
Chapter Three (§§8–15)	Sections A, B
PART ONE	Chapter 3
Chapter One (§§16–18)	Sections A, B
Chapter Two (§§19–23)	Chapter 4
Chapter Three (§§24–28)	Sections A, B
Chapter Four (§§29–36)	Section C
Chapter Five (§§37, 38)	C1–C4
PART TWO	Chapter 5
Chapter One (§§39–41)	Sections A–F
Chapter Two (§§42–44)	
Chapter Three (§§45–48)	
Chapter Four (§§49–61)	
Chapter Five (§§62,63)	
Chapter Six (§§64–76)	

of our fellow animals. As PRELIMINARY APPRAISAL and PART ONE say little directly about our fellow animals, a non-unitary understanding of *FCM* would delimit any interpretation of our fellow animals to the contents of PART TWO. Heidegger is clear that this is *not* the case. For example, near the end of *FCM* (p. 351) Heidegger says,

> The awakening of a fundamental attunement in the first part of the lecture course, and the return we have just undertaken from the logos-structure to the fundamental occurrence, both serve one purpose: *to prepare our entering into the occurrence of the prevailing of world.* … Awakening is a matter for each individual human being, not a matter of his or her good will or even skillfulness, but of his or her destiny.

In other words, profound boredom (= the topic of PART ONE) necessarily operates for each human *Dasein* within a specific context that is somehow related to Heidegger's comparison of humans and our fellow animals (= the topic of PART TWO). For Heidegger, elucidating this comparison in its profundity requires a philosophical approach, what he calls "comprehensive thinking" (= the topic of PRELIMINARY APPRAISAL). So, to understand why Heidegger is saying what he does about our fellow animals in PART TWO, we cannot ignore other parts of the book, even including sections of PART TWO (particularly, §§68–76) where animals are little mentioned. Simply stated, *FCM* must be read in its unity. This is not to say that all sections of *FCM* are equally relevant to understanding Heidegger's view of our fellow animals. PART TWO (Chapters Two–Five) remains Heidegger's clearest statement on this matter, so that part is emphasized in my discussion.

Withy (2013) has presented a compelling framework for understanding *FCM* in its unity, and her central arguments can be usefully abstracted. She argues that in *FCM* Heidegger aims to ready us (i.e., the students who are attending his lectures in 1929/30, and later, us the readers of the book) for the possibility of becoming authentic human beings, or better said, of becoming authentic human *Dasein*, through a sort of "activist metaphysics" that he introduces in PRELIMINARY APPRAISAL. Withy argues that *FCM* should not be read as a theoretical or scientific treatise. In particular, it should not be read as an analytic comparison of humans and our fellow animals but rather as an effort by Heidegger to guide us (i.e., human *Dasein*) toward a comprehensive understanding of what it means *for us* to have a world—not to be confused with a modern understanding of nature—and why such an awareness of world is vital. In *FCM*, Heidegger claims that "having a world" is restricted to human *Dasein*, even though we humans typically are unaware of this and instead falsely attribute "having a world" to our fellow animals as well, or so Heidegger argues in PART TWO. If we attribute the same "having a world" to our fellow animals as to ourselves, then we must, according to Heidegger, be misunderstanding our very own world. The task of clarifying why this is so—and jarring us into this alternative realization—requires philosophical thinking, but of a special sort. Heidegger says,

> Although we have spoken *of* philosophy, we have not yet spoken *from out of it*. We have indeed dealt *with* philosophy, but not taken action *within* philosophy itself. What is decisive, however, is that we emerge from this dealing with … and *take action within metaphysics itself*. This means nothing other than the fact that we must now really and properly question.
>
> (*FCM*: 57)

And on that same page, Heidegger continues,

> We have already indicated the questions we are posing: What is world? What is finitude? What is individuation? Ought we therefore to pose these questions and bring about a suitable attunement for them? If we were to proceed in this way, we would only fall back again onto the level which we wished to abandon and must abandon. We would have theoretical discussions (science) which we would then subsequently bring to fruition in terms of a worldview. The entire superficiality of philosophizing would thereby be restored. It is not a matter of developing these questions as theoretical and then bringing about an attunement *in addition* or *alongside*. On the contrary, we must first of all let these questions arise in their necessity and possibility from out of a fundamental attunement, and seek to preserve them in their independence and unambiguousness.

As a theoretical comparative work, Withy says *FCM* fails, but as a therapeutic and revelatory work, it succeeds. The terms "therapeutic" and "revelatory" can be easily misunderstood. They refer to Heidegger's thinking *ontologically*; that is, they refer to human *Dasein*'s very being and should not be thought as psychological terms. "Therapeutic" points toward projecting a *curative* possibility for the malady of humanity's inauthenticity,[2] and "revelatory" points toward the possibility of authenticity[3] as something that may be *given* to us by being itself. Withy's reading of *FCM* as one directed toward providing therapy and revelation for humans rather than some sort of objective naturalistic comparison between humans and our fellow animals makes sense in light of Heidegger's rejection of (object) naturalism and his claim of human *Dasein*'s uniqueness. Withy's reading of *FCM* as a unity of three parts also makes sense of Heidegger's capacious organizational skills in writing and lecturing.

Section B. Commentary on PRELIMINARY APPRAISAL: The Task of the Course and Its Fundamental Orientation, Starting with a General Elucidation of the Title of the Course

The text of PRELIMINARY APPRAISAL (§§1–15) occupies the first 16 percent of *FCM*, yet its contents are little mentioned by commentators when discussing Heidegger's views on our fellow animals (but see Withy 2013). Most commentators focus on the text's direct statements about animals (including humans) or on the relationship, or not, between PART ONE (on boredom) and PART

TWO (the comparative part). This could appear reasonable if PRELIMINARY APPRAISAL were simply an overview meant to entice the reader to read on, but it is not. So, "what is happening to us here?"[4]

PRELIMINARY APPRAISAL is a treatise on how to approach philosophizing, not in the sense of acting from within an established academic discipline but through a sense of being "grabbed" by what Heidegger calls a fundamental attunement of philosophy. Called *homesickness*, this attunement is not a psychological state but rather a moment in authentic human existence. It is a drivenness to be at home everywhere—within the whole or what Heidegger calls "world." Further, Heidegger says,

> this urge to be at home everywhere is in itself at the same time a seeking of those ways which open up the right path for such questions. For this, in turn, we require the hammer of conceptual comprehension, we require those concepts which can open such a path. We are dealing with a conceptual comprehension and with concepts of a *primordial* kind.
>
> (*FCM*: 6)

In other words, *FCM* is a philosophizing that is meant to make accessible to his students—and now to us the readers—an entryway to their/our own comprehensive questioning about *Dasein* and its world. To do this, we must be properly attuned, but to be attuned requires (proper) philosophizing. Heidegger's philosophizing entails using concepts that are not of current fashion or of historical precedent even though they remain grounded in their origin, where "grounded" does not mean "identical with" some previous concept but "possibilities emerging out of" in a manner that is true to the original phenomenon, namely world. We can extend Withy's (2013) unifying insight from PART ONE and PART TWO to PRELIMINARY ANALYSIS by saying that PRELIMINARY ANALYSIS lays out the means and terms by which unification of the remainder of the book takes place.

Heidegger's means and terms are not without bias. For example, the human is uniquely addressed as having full-blown language skills, thereby undercutting interest in the genesis of language and in shared communicative skills with our fellow animals (see Brandom 2000 for a concise acknowledgment of and justification for this move). Another bias is his introduction of important concepts in the early portions of the book that in later portions strongly influence the meaning of subsequently introduced concepts, and in particular, those concerning our fellow animals. At first blush (i.e., our commonplace understanding), some of these later-introduced concepts may not appear particularly unknowing or insensitive with regard to our fellow animals, but that changes once Heidegger interprets them in light of the earlier concepts. To put the issue more bluntly, it is as though Heidegger's original concepts already have embedded within them the later concepts, which are simply disclosed through philosophical analysis. So, it is essential that we briefly consider some of the key concepts introduced in PRELIMINARY APPRAISAL.

The bias issue aside, understanding Heidegger's language can still not be straightforward. For example, he attaches his fundamental philosophical concepts to ordinary words but now with meanings that are quite different.[5] Reading Heidegger without this awareness leads only to confusion. To confront this language barrier as quickly as possible, I comment below on six fundamental concepts that appear in PRELIMINARY APPRAISAL. My comments are brief and are meant to introduce and guide my analysis of *FCM*; they are not meant to be definitive. After all, Heidegger devotes entire books to some of these concepts, and their meanings vary over Heidegger's lifetime. Furthermore, for Heidegger philosophical concepts cannot by their very nature be defined either extensionally or intentionally, but rather require what he calls *formal indications*—contentless "descriptions" that are meant to point to and enable further, deeper questioning.

My comments on each of the philosophical concepts have three components. I start with one "ordinary, commonplace" understanding of the concept (component *a*). Then, I describe Heidegger's understanding of the same term in *FCM*—and sometimes in other publications of his—for humans (component *b*). Then, I indicate how Heidegger sees the concept relating to our fellow animals in *FCM* (component *c*).

1. Essence and existence. **(a)** In commonplace understanding—an understanding that is derived from Scholastic thought—essence refers to the quality whereby an entity is what it is ("what-being") in its necessary and permanent being, separate from the existence of that entity ("that-being") in its actuality. Relatedly, the essence of an entity is considered its "nature," as when one says, "It is in the nature of the human to have a soul" or "It is in the nature of the lion to be ferocious." **(b)** In PRELIMINARY APPRAISAL (p. 21), Heidegger says, "Thus man in the ground of his essence is someone in the grip of an attack, attacked by the fact 'that he is what he is,' and already caught up in all comprehending questioning." Relatedly in *Being and Time* (p. 67), he says, "The essence of Dasein lies in its existence." In humans, in other words, essence is manifest through the dynamic of existing, which is future oriented, and is to be distinguished from the actuality of being, which is present oriented. Elsewhere in *Being and Time* (p. 33), he says that existence is to be understood in terms of possibility, not actuality: "Dasein always understands itself in terms of its existence—in terms of a possibility of itself." And these possibilities must matter to *Dasein*; otherwise, existence becomes simply living in a future-less present. **(c)** The two terms—essence and existence—are to be thought in their unity, but since existence, as per Heidegger's usage, is restricted to humans (*Being and Time:* 67), the being of humans and that of our fellow animals must be separated by an abyss. Later in *FCM* (pp. 196, 282), Heidegger describes the essence of animality as captivation and by his thesis: "the animal is poor in world." The word "poor" indicates that as humans we might imagine a simulacrum as to how our fellow animals live in a world, but that for the animals themselves, there simply is no such world. Our fellow animals, according to Heidegger, are distinguishable from the

stone largely by the imaginings of humans ("the stone is worldless" [*FCM*: 196]). Their essence is locked into a present, while our essence is projected into a future.

2. World. **(a)** In commonplace understanding, the world "signifies the totality of those entities which can be present-at-hand within the world" (*Being and Time*: 93), including humans and our fellow animals. **(b)** In PRELIMINARY APPRAISAL (p. 5), Heidegger says, "to be at home everywhere means to be at once and at all times within the whole. We name this '*within the whole*' and its character of wholeness the *world*." World is a meaningful prefiguration of mutually referential contexts and is not something present-at-hand. The prefiguration—and particularly its manifesting—as opposed to the referential contexts themselves, point toward the ontological. In *Being and Time* (p. 93), he calls his somewhat different ontological understanding "worlding." After *FCM*, he modifies his ontological understanding again, but still refers to "world." **(c)** Later in *FCM* (p. 196), Heidegger says, "the animal is poor in world." Basically, Heidegger concludes that only human *Dasein* has a world; our fellow animals have no world.
3. Finitude. **(a)** In commonplace understanding, finitude identifies a limitation on human ability relative to that of an infinite God, but it is also applied to any living being. **(b)** In PRELIMINARY APPRAISAL (p. 6), Heidegger says finitude represents a fundamental way of being for humans, tethered in an "oscillating" between beings and being. Far from a limitation, our finitude enables human freedom by urging us toward being (Withy 2013). Finitude is a dynamism which makes authentic human *Dasein* possible. Part of *Dasein*'s dynamism is its language, "that belongs to the essence of man in his finitude. To imagine a god expressing himself in speech is utterly meaningless" (*FCM*: 238). In other words, finitude is our fate because we have a relationship to being but are not being itself. Our finitude is the expression of our very essence. More straightforwardly, we might say that *Dasein*'s attunement and projection, whereby a world appears to *Dasein* in a meaningful way, is our finitude. **(c)** Later in *FCM* (p. 238), Heidegger implicitly contrasts human finitude with a way of being of our fellow animals:

> For in its behavior—as that which the animal is capable of—in this driven performing the animal does not drive itself away from itself. Rather the animal *is* in precisely such a way that it intrinsically retains itself and is intrinsically absorbed in itself. Behavior and its forms are not something which radiate outward and allow the animal to run ahead along certain paths. Rather *behavior* is precisely an *intrinsic retention* and *intrinsic absorption*, although no reflection is involved.

Stuck in absorbed suspension and without language, our fellow animals, according to Heidegger, simply cannot be thought of as finite, for to be finite is to project possibilities through freedom.

4. Individuation (which leads to the phenomenon of *solitude*, a word found in *FCM's* subtitle). **(a)** In commonplace understanding, individuation is a separation, a distinguishing from others, or a separation into a unit. It can be applied to any living being. **(b)** In PRELIMINARY APPRAISAL (p. 6), Heidegger says that authentic human *Dasein* is necessarily taken up by individual human *Dasein*. Becoming authentic is not something we know but something we do. Becoming authentic requires being oneself, taking responsibility for the truthfulness of one's comportments toward other entities. Thus, being an individual (individuation) is an essential aspect of authenticity, along with having a world and being finite. **(c)** Later in *FCM*, we see no reference to individuation of our fellow animals. Indeed, animals as a collective (of course, excluding humans) are the frame of Heidegger's talk.

> The animal is poor in world. … We see at once that in fact our thesis does not tell us something merely about insects or merely about mammals, since it also includes, for example, non-articulated creatures, unicellular animals like amoebae, infusoria, sea urchins and the like—*all* animals, *every* animal. … this thesis is meant to say something about animality as such, something about the essence of the animal: it is a *statement of essence*. It is not a statement of essence simply because it holds true for all animals and not merely for some of them. Rather, it is the other way around: it holds true for all animals because it is a statement of essence.
>
> (*FCM*: 186)

In summary, we can say that individuation is centrally related to being an authentic human *Dasein* but that such considerations for our fellow animals are inconceivable in Heidegger's understanding.

5. *Dasein and Da-sein.* **(a)** In commonplace understanding, *Dasein* is equivalent to the human or to human consciousness. "*Dasein*" is literally translated as "being there/here" and is frequently taken to indicate "being present" in a naturalistic sense.[6] Heidegger's distinction between *Dasein* and *Da-sein* is ignored. **(b)** In PRELIMINARY APPRAISAL, *Dasein* is a being who is distinguished from non-*Dasein* beings by having a distinct way of being, a way whose meaning can be at issue for *Dasein*. Said more straightforwardly, *Dasein* is a unique sense-maker of her world and of herself. *Dasein* is never to be thought as present-at-hand alongside other material things. As Heidegger says, "Dasein is not something that one takes for a drive in the car as it were, but something that man must specifically take upon himself" (*FCM*: 165). From this quote we might think (as Fried and Polt do; see Heidegger 2014: xi) of *Dasein* "as a condition into which human beings enter, either individually or collectively, at a historical juncture when the Being of beings becomes an issue for them, or Being as the event of meaningful disclosure takes place for them." As being-in-the-world, *Dasein*—also called "human *Dasein*" or sometimes simply "human" in *FCM*—has factical determinations both as "self" and as "world." But only in a derivative sense can human *Dasein* be

thought of in isolation, either as an autonomous self or as cultural norms. In *Being and Time*, Heidegger says that human *Dasein* can have an authentic relation to its being, in which case it is a "true self," or *Dasein* can have an inauthentic relation, in which case it is an "anyone-self." Anyone-self can be thought of in grammatical terms as a "distributive singular term" (per Wilfrid Sellars; see Brandom 2015: 247). In summary, *Dasein* is Heidegger's fundamental way of thinking human being without reducing the human to rational being (e.g., as found in Enlightenment thinking), conscious/sentient/sapient being (e.g., in psychological thinking), or created being (e.g., in religious thinking).

In response to widespread misunderstanding of his *Dasein* thinking, Heidegger later coins the term "*Da-sein*" to emphasize that *Dasein*, and authentic *Dasein* in particular, is never to be thought anthropologically. For the same reason, *Dasein*'s way of being, called "existence," is later changed to "ek-sistence." Unlike its naturalistic counterpart, Heidegger's existence (and ek-sistence) emphasizes dynamic possibility over static actuality. As ek-sistence, *Da-sein* is individualized and endures, and is accountable to him/her-self and others to preserve the "there"/"here" (*Da*), namely, the "site" within which being and beings appear as meaningful. In a parallel manner, that which individuates and makes entities salient for *Da-sein* is now referred to as "be-ing," replacing "being." Unlike *Dasein*, it is less clear that *Da-sein* can refer to all humans.

(c) Later in *FCM*, when Heidegger is directly talking about our fellow animals, they are never referred to as *Dasein*. "*Dasein*" is a word that distinguishes the human from our fellow animals. Such a dichotomy pre-establishes Heidegger's conclusion that the being of *Dasein* and that of our fellow animals ("animality") are incomparable, that is, separated by an abyss.

Beyond these general comments about *Dasein*, Heidegger describes numerous aspects of *Dasein* that serve as formal indications of *Dasein*'s own ontological structure. He calls these "existentialia" (singular: existentiale), and these illuminate how *Dasein*—and *Dasein* only—can be exposed to the unconcealedness of beings as such (List 1.1). Existentialia relate to *Dasein*'s being, as opposed to characters, or categorials, that relate to non-*Dasein* entities (e.g., to equipment and our fellow animals). Existentialia can also be contrasted to *Dasein*'s "existentiells"; the latter relate *ontically* to some aspect of one's own *Dasein* as an entity. For example, a biologist's understanding of his/her own role as a biologist is existentiell, although his/her understanding of other colleagues' roles is neither existentiell nor existential; it is simply ontic, as is any naturalistic understanding of the (human or otherwise) organism. Heidegger is not particularly interested in the ontic or the existentiell except as they illuminate *Dasein*'s (ontological) existence. As these distinctions hopefully make clear, Heidegger's writing about *Dasein*, including in *FCM*, should not be confused with writing about the human as understood in everyday language, but it is also clear that keeping humans and our fellow animals separate is built into Heidegger's word choices

List 1.1. *Existentialia of human* Dasein*

- affectedness / *Betroffenheit* (Schalow: 43)
- being-alongside entities within-the-world (BT: 81, 264)
- being-in (BT: 79, 82)
- being-possible (BT: 292)
- being-toward-death (Schalow: 235)
- being-with/*Mitsein* (BT:156, Dreyfus: 20)
- concern (BT: 83, 158)
- de-severance (BT: 139)
- directionality (BT: 138, 143)
- disclosedness (Dahlstrom: 71)
- discourse/talk (BT: 203–4, 209, 266)
- disposedness/state-of-mind/*Befindlichkeit* (Schalow: 108; BT: 172–3, 203)
- existence (Schalow: 120)
- facticity/thrownness (Schalow: 120, 129; BT: 174; Dreyfus: 20)
- fallenness (Schalow: 120, 130; BT: 220–1, 264)
- having a conscience (Schalow: 235)
- "in each case mine" (Blattner: 33)
- making room (Schalow: 224, BT: 146)
- meaning/sense (Schalow: 226; BT: 193, 370–1)
- possibility (Dahlstrom: 70; BT: 143 [footnote])
- potentiality-for-being-a-whole (BT: 277, 311, 357)
- potentiality-for-being-in-the-world (BT: 342)
- potentiality-for-being-one's [its]-self (BT: 312, 314, 351, 369)
- projection (BT: 185, 264)
- resoluteness (BT: 343–5)
- solicitude (BT: 158)
- "the they" (Schalow: 351, BT: 167)
- truth/primordial truth (BT: 269, 343)
- understanding (BT: 182–4, 203, 385)
- worldhood (BT: 92)

* This list is not meant to be all-inclusive. Individual existentialia are followed by references and page numbers (in parentheses). Abbreviations: "Blattner": Blattner 2006; "BT": Heidegger 1962; "Dahlstrom": Dahlstrom 2013; "Dreyfus": Dreyfus 1991; "Schalow": Schalow 2020.

....

6. *Philosophy.* **(a)** In commonplace understanding, philosophy can be thought of as one worldview or another (e.g., Aristotle's or Kant's philosophy), or as that which provides the most basic or general concepts for the so-called positive sciences (i.e., those directed toward understanding some entity; for example, biology is directed toward *life* while theology is directed toward

God). Philosophy is a human endeavor that is largely about humans. **(b)** In PRELIMINARY APPRAISAL, Heidegger begins *FCM* (pp. 1, 2) with the statement, "Philosophy neither science, nor the proclamation of a worldview" and concludes §1a by saying, "philosophy is something that *stands on its own*, something *ultimate*." He elaborates: "Philosophizing stands on its own as the fundamental occurrence in Dasein" (*FCM:* 21), and "Metaphysics is a questioning in which we inquire into beings as a whole, and inquire in such a way that in so doing we ourselves, the questioners, are thereby also included in the question, placed into question" (*FCM:* 9). **(c)** Later in *FCM*, Heidegger identifies a problem with these understandings of philosophizing—they all too easily place human *Dasein* and its being at the center of the investigation largely to the exclusion of the rest of the world. We might ask—although Heidegger clearly does not—whether animals too can relate to philosophizing, or to thinking more generally. In PART TWO, where our fellow animals are explicitly considered, it is clear that the science of ethology plays some role in Heidegger's thinking about the being of animality, but "translating" Heidegger's understanding of animal behavior into animal thinking remains far-fetched.

The latter three-fifths of PRELIMINARY APPRAISAL (= the entirety of Chapter Three) presents an etymological analysis of the word "metaphysics." One root of the word lies in the Greek word *phusis*, commonly understood as "nature" or "object of natural science" but which Heidegger translates in its original (i.e., in pre-Socratic times) meaning as "the self-forming prevailing of beings as a whole" (*FCM:* 25). This prevailing of beings as a whole—similar to Heidegger's term "presencing"—is logically prior to any modern understanding of nature or history. For example, happenings of significance throughout a human life are not, according to Heidegger, fundamentally biological events but rather are the prevailing of beings, understood in the context of their history and fate.[7] "Prevailing" indicates presence, but that requires a prior absence, or concealment (Withy 2022). It is the *logos* (according to Heidegger's explication of Greek philosophy) that "takes this prevailing from concealment." Later in *FCM*, Heidegger will connect *logos* to "discourse" (which is not to be equated with speaking), claiming that it is not available to our fellow animals.

Heidegger articulates two aspects of pre-Socratic philosophical thinking about *phusis*. The first aspect addresses beings as a whole, or that which prevails, and derivatively addresses regions of beings, for example, those beings confined to specific fields of natural science. The second aspect—called *ousia* by Aristotle—addresses the essence or nature of beings as such—"the prevailing of what prevails" or "that which determines whatever prevails as a being, that which makes beings beings."

After Aristotle, Heidegger argues that philosophy loses its unitary understanding of *phusis* and eventually fragments into distinct disciplines that largely map to *phusis*'s first aspect only. One such discipline is ethics, whereby we humans question our self-conduct as a being—a being distinct from nature, understood naturalistically—but not as being (or *is-ing* to use

Withy [2017]'s pithy term). The failure of ethics to explicitly incorporate both understandings of *phusis* makes it a derivative "science." The same applies to the so-called natural sciences, including—most relevant to my book—ethology and evolutionary biology.

Heidegger concludes by returning to the importance of *doing* metaphysics, of being gripped in a comprehensive questioning through an inquiry into beings as a whole and in their being. This "as a whole" adumbrates what Heidegger refers to as *world*, but what is this? And what is being?: Not a thing, and yet not *no*thing. As finite creatures, our access to being is through beings. While beings can be accessed through science (theoretical inquiry), being cannot because being is not a being. Comprehensive questioning of being and world must arise, in Heidegger's then-contemporary understanding, in its "necessity and possibility from out of a fundamental attunement" (*FCM*: 57). In the next section, we consider what Heidegger has to say about a particular fundamental attunement. But that attunement turns out to be dependent on the concepts developed in *Preliminary Appraisal*, where we must wonder whether our fellow animals have been given proper account under the guise of methodological questions about philosophy and etymology.

Section C. Commentary on PART ONE: Awakening a Fundamental Attunement in Our Philosophizing

In both *FCM* and *Being and Time*, Heidegger tells us in no uncertain terms that we are seldom authentic *Dasein*. However, the two accounts differ as to how we can escape from an impoverished state of inauthenticity (Withy 2013). In *Being and Time*, the challenge is to overcome our absorption in everyday *Dasein*, to turn away from what Heidegger calls *das Man*. To do this, we must be in a fundamental attunement of anxiety as it relates to being-in-the-world, and particularly to our own death. We must confront the *nothing* of our existence in order to become a *self*. In *FCM*, the challenge is to overcome our absorption in (the world of) entities, where all things are present-at-hand and where we understand the being of all beings as equivalent—most relevantly here, that we misguidedly view ourselves as just another animal. To see our own inauthenticity, Heidegger says we must be in the fundamental attunement of profound boredom, which leads to a *moment of vision*—a reinterpretation of our own *Dasein* and of the world and its entities in their truth. To become a self, we must confront beings in their essence and as a whole. The moment of vision opens us up to the possibility of doing comprehensive thinking. Assisting us in this task is the aim of PART ONE of *FCM* (§§16–38). It is important to emphasize that Heidegger presents this discussion of profound boredom before presenting the direct contrast between humans and other animals and between human *being* and animal *being* in PART TWO. Presumably, this is so that we who become profoundly bored will then be able to "appropriately" interpret the scientific record presented in PART TWO.

....

Table 1.1 Levels of Boredom as Described in *The Fundamental Concepts of Metaphysics*

phenomenal characterization	**1st structural moment (generically, a withdrawing/withdrawal: "being left empty")**	**2nd structural moment (generically, a revealing/revelation: "being held in limbo" *)**
LEVEL I: "bored *by* something" (*by* a determinate thing or situation)	being left empty by those beings surrounding us in the situation that we find ourself	being held in limbo *by* the dragging of time that we somehow need
LEVEL II: "bored *with* I know not what" (*with* some indeterminate thing or situation/with the passing of time itself)	being left empty to whatever is going on in a self-forming emptiness ("obstructive casualness"); leaving your authentic self behind	not being released from our time as being held in limbo as time stands still in its standing "now" (time does not bind us to itself, to "the flow of enduring")
LEVEL III ("profound boredom proper"): "it is boring *for* one" (*for* an undifferentiated no one who is bored *by everything and everyone, including ourselves*)	being delivered over to being's telling refusal of themselves as a whole (i.e., all individual entities lose their significance) through the entrancement of *Dasein*'s temporal horizon (leading the past, present, and future to merge into an undifferentiated temporality), and leading to an absence of oppressiveness as a whole (the absence of which is what actually oppresses us and leaves us profoundly empty, i.e., of one's possibilities), although *we are nevertheless compelled to listen to what this boredom has to tell us*	although empty and held in limbo, simultaneously and paradoxically in a telling announcement of *Dasein*'s possibilities left unexploited, through the rupture of the entrancement of time (i.e., through the expansion of the temporal horizon), being impelled toward the extremity of originary making-possible of *Dasein* as such, *Dasein*'s resolute self-disclosure (self-unconcealing) as the moment of vision, as the fundamental possibility of *Dasein*'s existence proper. [alternate phrasing: the absence of passing time makes its boredom overpowering, guiding *Dasein* to its moment of vision (*Augenblick*) in a totalizing "event" (*das Ereignis*)]
LEVEL IIIa ("contemporary form of profound boredom," or simply "contemporary boredom")	being delivered over to beings' telling refusal of themselves as a whole in the telling absence of any oppressiveness	*the absence of a telling announcement* resulting in being left empty by *not* being held in the limbo of profound boredom proper

*Agamben (2004) translates this phrase as "being held in suspense."

To make profound boredom comprehensible to others, Heidegger conceptually separates *Dasein*'s attunements as boredom into levels of increasing essentiality that parallel his earlier discursive transition from attunement to profound attunement.[8] There are three levels (*bored by*, *bored with*, and *profound boredom*) and one derivative form of profound boredom (*contemporary boredom*). Each has a distinct structure (Table 1.1). It is important to emphasize that these levels have a unity which becomes most apparent through the way in which time (temporality; Blattner 2005) adjudicates the particular phenomenon—*dragging*, *standing*, and *the absence of passing time*. It is also important to emphasize that Heidegger's task for the students in his class is not simply to provide an interpretation of boredom but rather to assist their becoming open to their own profound boredom. Heidegger says, "Attunements are the fundamental ways in which we find ourselves disposed in such and such a way. Attunements are the 'how' according to which one is in such and such a way" (*FCM*: 67). So, attunements according to Heidegger are not psychological states but the condition of the possibility of such states; they are pre-reflective ways in which we make sense of the world. They can disclose both ontological and ontical structures (*Befindlichkeit* and *Stimmung*, respectively). Indeed, *Dasein* is always attuned in some way or other, although we are seldom aware of this. Attunement is Heidegger's way of moving our orientation on the world from a psychological state to something that is neither exclusively interior nor exterior to our individual *Dasein*, but rather is necessarily already *in* the world.

The first two levels of boredom are clearly enough related to our ordinary understanding of boredom, our ordinary ways of being bored in time with some determinate (level I) or indeterminate (level II) thing or situation. Contra Heidegger, many contemporary observers (myself included) of animal behavior would argue that these same sorts of boredom are also open to at least some other animals, despite Heidegger's conceptually abstract way of framing. But regardless, it is the third level of boredom, when time stops being oppressive in the *telling refusal*, that most interests Heidegger and which he clearly states is open to human *Dasein* only. What is refused is *Dasein*'s own continuing immersion, or absorption by, the surrounding entities. Instead, while these entities are still there, they become undifferentiated as a whole and, thus, are no longer of interest. Another way of framing "are no longer of interest" is to say that different entities become indistinguishable. In particular, some humans—here, think biologically—see themselves as fundamentally indistinguishable from our fellow animals. If *Dasein* stops there, we are left with what Heidegger calls contemporary boredom, which is the condition of most contemporary humans. However, this removal from absorption in individual entities as meaningful can also open us to "seeing" new possibilities through a *telling announcement*—a revelation—that brings us face-to-face with beings *and* being. The announcement is not a demand to be authentic as such but rather a change in our very openness to hearing this demand and burden of self-transformation. The result is a "timeless" reinterpretation of one's world in a way that is authentic to one's very own *Dasein*. This reinterpretation gives rise to an "oscillation" between a new, authentic understanding of beings and being; we are now attuned in the attunement of profound boredom. We now "see" the world as meaningful in the essentiality of our finiteness (Withy 2013). Importantly, we now see ourselves as radically distinct from other entities, including our fellow animals.

I conclude this brief discussion on attunement found in PART ONE by considering how Heidegger is leading us toward his understanding of our fellow animals through his preparatory consideration of profound boredom. According to Heidegger, unless we humans reorient our understanding of the world, we continue to see it largely in naturalistic terms with its exclusive focus on beings. By contrast, when profoundly bored, we are open to new possibilities for ourselves. But, as Heidegger says, learning about boredom is not enough to become bored; indeed, knowing *about* boredom may play no role whatsoever. Instead, we need to be grabbed by profound boredom itself. Once so grabbed, *Dasein* becomes open to seeing—through Heidegger's originary comparison between ourselves and our fellow animals presented in PART TWO—that we are not animals. For Heidegger, profound boredom opens up the ontological abyss between human *Dasein* and animals.

Section D. Commentary on PART TWO: Really Asking the Metaphysical Questions to Be Developed from the Fundamental Attunement of Profound Boredom. The Question: What Is World?

While PART TWO (§§39–76) of *FCM* is commonly considered to be the comparative part, it is clear from its title and introductory first chapter that metaphysical questioning of world and our own *Dasein* in the fundamental attunement of profound boredom remains a central concern of Heidegger. This reinforces the thought that *FCM* is not primarily a book about our fellow animals.

Chapter Two of PART TWO is concerned with further pre-structuring our way of thinking about world. In other near-contemporaneous works, Heidegger approaches the question of world through a historical, etymological approach (e.g., "On the Essence of Ground"; Heidegger 1998c) in which humans typically (according to Heidegger) are considered as master (of non-*Dasein*) and servant (to God) in a world that is set off against God. In *Being and Time*, Heidegger presents a second way of considering world, through the everyday understanding of human *Dasein*. Both attempts emphasize the tight linkage between world and humans, so we should not be surprised that his third way to explicating world—the comparative approach of *FCM*—is likewise principally directed toward understanding human *Dasein*. He says, "Man has world. But then what about the other beings …?" To which he responds with his well-known distinctions, or theses, which "immediately manifest themselves," namely, "(1) the stone (material object) is *worldless*; (2) the animal is *poor in world*; (3) man is *world-forming*" (*FCM*: 177). We might wonder why Heidegger articulates the world into these three realms—stone, animal, human—in what amounts to a quite substantive claim. But in my discussion of PRELIMINARY APPRAISAL and PART ONE, I argued that Heidegger had already laid the conceptual framework to support the division between "animal" and "man." These theses, then, incorporate that framework in what is Heidegger's next big and most explicit step toward separating human *Dasein*'s way of being from that of all other entities.

According to Heidegger, before starting a comparative analysis of *Dasein* and our fellow animals, we must know their essences, or their being. That is the philosophical task that Heidegger sets out for himself in PART TWO and toward which he aims to inspire his readers. Doing this does not simply entail making a comparison of different beings (i.e., *Dasein* and our fellow animals), but rather making an interpretation of their (way of) being (i.e., of *Dasein*'s existence and of our fellow animal's animality). In making comparisons (and interpretations), Heidegger says that the materiality (e.g., embodiment) of beings is of secondary concern. By contrast, embodiment is a primary concern for biologists. But it is with behavior as such where the interests of Heidegger and biologists partially overlap, largely because Heidegger considers behavior to be a unified feature of the animal in a manner that parallels his understanding in *Being and Time* of comportment of the human. This overlap is interesting because behavior (and comportment) is not being; it is an entity, just not a material one.

So, how does Heidegger make the conceptual leap from behavior (and comportment) to essence? The key, he says, is to do philosophy. We must know *what* the essence of life in general (for Heidegger, animals) is and *how* animals are accessible as such. As concerns the "what," it is behavior—versus more reductionist accounts—that provides ontological access to the animal. As concerns the "how," Heidegger says the philosopher must move "in a circle" by presupposing the essence, seeing where it leads, and then deciding whether or not to re-enter the circle with a revised presupposition. The metaphor of entering a hermeneutic circle sounds reasonable as long as choosing different entry points does not stably lead (i.e., after multiple iterations) to distinct understandings. If it does and if there is no basis outside the circle itself for deciding where to enter, then one might feel compelled to acknowledge the possibility of a plurality of understandings. Heidegger further claims that his hermeneutic approach distinguishes philosophy from science, but this is only partially correct. Defeasibility is a related core principle of (ontical) science, and this indicates that scientific questioning and interpretation remain ongoing, analogous to re-entering Heidegger's (ontological) hermeneutic circle.

The comparative analysis of *FCM* begins in earnest in Chapter Three of PART TWO, slightly more than halfway through the entire book. At the outset, Heidegger reminds us that "What is world?" is our question (*FCM*: 185). Expanding on this, he says that the perspective from which this question is asked "is the *specific relation* that stone, animal, and man in each case has toward *world*" (*FCM*: 185). But it is the relationship between our fellow animals and their world where he says we ought to begin. Heidegger might explain that starting with rocks would seem ill-conceived, as they *obviously* have no world, and thus their analysis would seem to have little to contribute to our understanding of *Dasein*. And, conversely, starting with humans would make what follows—the animal part—largely incidental. The point of comparison would be reduced because the main (human) part would already have been expounded, leaving little reason for comparison. So, our fellow animals are featured prominently in this portion of *FCM*.

Heidegger again cautions against considering "stone, animal, and man" as of the same order, contra a reductionist orientation in which all beings share and are understood through their material aspects and interactions, or relations. He further cautions about relation-talk itself because it can easily lead to a consideration of everything as present-at-hand (*FCM*: 293). Relation-talk is useful only to the extent that it leads us to the proper and peculiar task of philosophical thinking.

Chapters Three–Five of PART TWO are largely devoted to exploring the thesis "the animal is poor in world," but there is also an interesting subtheme that concerns the nature of biological research. To evolutionary biologists (such as myself), it must seem peculiar that Heidegger says his second thesis applies to *all* of our fellow animals.[9]

> This thesis is meant to say something about animality as such, something about the essence of the animal: it is a statement of essence. It is not a statement of essence simply because it holds true for all animals; rather, it is the other way around: it holds true for all animals because it is a statement of essence.
>
> (*FCM*: 186)

In contrast to his claim of a unitary understanding of the essence of our fellow animals, Heidegger says that the essence of *Dasein* itself must remain open to the possibility of a pluralistic understanding based on its distinct ways of being, for example, the sciences, language, and art. Of course, our fellow animals too have distinct ontical ways, as Heidegger acknowledges, but these differences do not, for him, undercut his presumption that their way of being is unitary.

Additionally, Heidegger says,

> But precisely because zoology deals with animals this proposition [that is, Heidegger's thesis about animals] cannot be a result of zoological investigation; rather, it must be its presupposition.
>
> (*FCM*: 187)

Here, he is saying that the ontical endeavors of zoologists presume, and do not test, their own ontological understanding of animal being.

Heidegger says that the being of all of our fellow animals can be meaningfully discussed as privative on *Dasein*. "Privative" in this sense means that our fellow animals can be understood, although only analogically, as lacking something present in human *Dasein*—namely, the human essence. The converse situation—that *Dasein* can be understood as animal plus something else—is excluded because having two essences makes no sense. Of course, Heidegger does not think that we can remove part of the (unitary) human essence to simulate the animal essence either; hence, Heidegger's resort to analogy. But still, should Heidegger's explanation as to why the analogy is allowed in only one direction be thought of as convincing?

It is easy to misunderstand Heidegger's thesis—"the animal is poor in world"—as one that is derivable from zoology. But Heidegger is clear that the thesis is first-and-foremost a metaphysical (ontological) one. He says that the thesis "requires a specific orientation toward zoology and biology in general, and yet it is not through them that its truth is to be determined" (*FCM*: 187). To understand the thesis, one must first enter the hermeneutic circle, where ambiguity is unavoidable (see *FCM*, §45a), and one must search for the animal's essence through philosophical thinking, while still using the "hints" provided by zoology. And even though these hints may appear ontically unambiguous—or so biologists might claim—their metaphysical interpretation, which necessarily underlies the ontical, can never be unambiguous, thereby putting ontical claims themselves into question as well. For Heidegger, there is no alternative, given that we must presuppose an animal's essence prior to any investigation. But even accepting this, it is still not obvious why Heidegger claims that his particular thesis must be the presupposition of zoology rather than just one possibility. Concerning this point, Heidegger has no decisive argument, but I think that his interpretation of animal behavior later in PART TWO of *FCM* is meant to persuade us that this is indeed the case.

In §45b ("The relation of our philosophical questioning to zoology and biology"), Heidegger claims that contemporary biology is in a metaphysical crisis as it tries to resist the materialist presumptions of physics and chemistry while also avoiding the emptiness of vitalism. He says that biologists typically avoid this crisis by being preoccupied with "practical and technical serviceability" (*FCM*: 191)—in modern terms, with procedures and practical goals—instead of seeing biology as but one possibility of human *Dasein*. Were Heidegger alive today, he might say that while physics and chemistry have become intercalated into the deepest recesses of modern biology, and while this has resulted in many new and exciting avenues of scientific research, the effect has been to hide our own *Dasein* even more than before, thus leading to a crisis that many scientists are completely unaware of. This "hiding" takes the form of a determinism versus the free possibility of human *Dasein*'s existence. Indeed, this theme of the danger of technological thinking—and its paucity of ambiguity about possibility—becomes increasingly central to Heidegger's thought after *FCM* (e.g., Heidegger 2013e).

Heidegger concludes §45 with the claim that humans have erred to the extent that they understand the essential nature of life (i.e., the being of animals and plants) from the perspective of human existence, from the perspective of materiality (e.g., a typical scientific view), or as something intermediate, which he calls "the biological worldview." So, we want to understand how Heidegger understands "life."

The opening of §46 announces how biological research fits into Heidegger's philosophical analysis.

> If we now return … to our thesis that the animal is poor in world, we must admit that recent research in biology, provided that we are capable of interpreting it in a philosophical way, strongly suggests the possibility of illustrating this thesis directly.
>
> (*FCM*: 192)

What does Heidegger mean here by "directly"? I believe he means that if a certain ontological interpretation is brought to bear on relevant biological phenomena, then the thesis can be confirmed. But why that interpretation and not another, for example, that of a research biologist? Had the thesis been presented as "the animal *in its essence* is poor in world," some clarity would have been provided. For Heidegger, there would have been no meaningful difference from the original, but of course biologists (and most other readers) might well have brought a different understanding to his claim. They would have more quickly realized that the terms Heidegger uses such as "animal," "essence," "poor," and "world" were all specialized philosophical terms, not biological ones. But even within Heidegger's frame, how can "recent research in biology" illustrate this thesis "directly"? After all, entering the hermeneutic circle necessarily entails ambiguity. Heidegger says, "The proposition [thesis] ... requires a specific orientation toward zoology and biology in general, and yet it is not through them [that is, zoology and biology] that its truth is to be determined" (*FCM*: 187). Again, biologists would be confused with Heidegger's usage of "truth" as unconcealedness (*aletheia*) rather than as correctness. For Heidegger, *aletheia* is prior to and makes possible the derivative sense of truth as correctness. So, in this case, Heidegger's truth is meant to reveal the animal's essence, not some scientific fact about the animal. But still, this does not clarify the relationship between the essence and the phenomena, and that is where Heidegger leaves us.

The thesis "the animal is poor in world" does not immediately illuminate itself, and §46 is meant to turn aside the "obvious" naturalistic (mis)interpretation. In the naturalistic scenario, "poor" refers to less of something, as opposed to more. We could, for example, easily imagine that the world of the human is enriched in its ability to penetrate that world relative to our fellow animals, but that is not what Heidegger is talking about. To illustrate this, he describes a worker bee who frequents blossoms for obtaining nectar but "does not know the stamens of these blossoms *as* stamens" (*FCM*: 193). By contrast, "We can characterize the relation man possesses to the world by referring to the extendability of everything that he relates to. This is why we speak of man as world-forming" (*FCM*: 193). But the implied idea that "world initially signifies the sum total of beings accessible to man or animals alike, variable as it is in range and depth of penetrability" (*FCM*: 193) with their corresponding different degrees of value ("poor in world" versus "world-forming") is not what Heidegger has in mind. In particular, this faulty explanation suffers from the naturalism that makes the worlds of our fellow animals and humans comparable, differing only quantitatively. It also suffers from human hubris. Heidegger famously says,

> However ready we are to rank man as a higher being with respect to the animal, such an assessment is deeply questionable, especially when we consider that man can sink lower than any animal. No animals can become depraved in the same way. ... Is the essence of man higher than the essence of the animal? All this is questionable even as a question.
>
> (*FCM*: 194)

Likewise, Heidegger says that it is a mistake to consider that "amoebae or infusoria are more imperfect or incomplete animals than elephants or apes. Every animal and every species of animal as such is just as perfect and complete as any other" (*FCM*: 194).

In §47, Heidegger compares our fellow animal's poverty in the world with the worldlessness of the stone, provisionally concluding that to have a world is for there to be beings that are accessible in such a way that they can be dealt with. The stone lacks all accessibility to beings. Heidegger states this as a bald fact. But the lizard (to take Heidegger's example) has a relation to the rock on which it suns itself, even if it does not (according to Heidegger) relate to the rock *as* a rock, or even to the rock as satisfying its immediate need. By contrast, the human does both. This difference cannot simply be due to a unique ability by humans to form concepts, since Heidegger is clear in *Being and Time* that the "existential-hermeneutic as" is prior to formal concept formation, although its relationship to language-ability as such is ambiguous in Heidegger's writings. But then, what happens to the lizard in its encounter with the rock? Heidegger says, "On the contrary, the lizard has its own relation to the rock, to the sun, and to a host of other things" (*FCM*: 198). What does "has its own relation" actually mean? Heidegger's response comes later in the book, but the short response is that any such relation is only from the perspective of the human observer. Heidegger generalizes from the lizard to all of our fellow animals and expands the list of immediate needs to "its sources of nourishment, its prey, its enemies, its sexual mates, and so on" (*FCM*: 198). Heidegger says,

> The animal's way of being, which we call "life", is not without access to what is around it and about it, to that amongst which it appears as a living being. It is because of this that the claim arises that the animal has an environmental world of its own within which it moves. Throughout the course of its life the animal is confined to its environmental world, immured as it were within a fixed sphere that is incapable of further expansion or contraction. ... Thus absolute deprivation of world does not belong to the animal after all.
>
> (*FCM*: 198–9)

There is a sense in which our fellow animals have a world—they have access to "an environmental world of its own," unlike the stone. But there is also a sense that they are deprived of a world—their "environmental world" (whatever that may be) is not the world that is accessible to humans as a world. So, something remains to be clarified about world, as Heidegger concludes at the end of Chapter Three.

Heidegger turns in Chapter Four (the second longest in the book) to clarify the sort of access that our fellow animals actually have. In that chapter, he talks about the essence of the animal, life in general, and organism. By now, it should be obvious that Heidegger is not talking about ontical distinctions among animal, life, and organism but about their *essences*, which are outside the investigative ken of biologists' investigations.

Heidegger begins §49 with the idea that the ability to access entities beyond oneself is related in some way to being able to *transpose* oneself ("self-transposition") into other beings, and then asks how this stands with humans and with our fellow animals. As regards the human, Heidegger asks three questions: "[1] Can we transpose ourselves into an animal? [2] Can we transpose ourselves into a stone? [3] Can we transpose ourselves into another human being?" (*FCM*: 202). As regards the first question, Heidegger reasonably rejects both actual transference and "as-if" or thought-experiment sort of transference. Instead, he says that transference is a "going along with another being," and that this "going along with" is unrelated to any empathy. To Heidegger, empathy suggests "an isolated being existing for himself" (*FCM*: 207), a kind of internalism or Cartesianism that is opposed to being-in-the-world. Rather, transference is a human awareness of the *ways* in which an animal goes about its life—seeing, hearing, seizing, evading, building. Whether or not we are successful in these factical tasks, they always remain possibilities. We think this is so because, Heidegger says, we mistakenly do not question the ability of our fellow animals to do the same. The principal window into this going-along-with is ethology, although the linkage between the ontical (ethological) and the ontological (metaphysical) remains obscure. Somehow, the way of life of our fellow animals as manifest in their behavior becomes accessible to ontological analysis.

All this is very interesting, reasonable, and typically Heideggerian. What Heidegger has done here is to reframe a commonplace but misguided way of thinking about transposition as a mentalistic affair. Instead, he says that transposition necessarily occurs in the world and cannot be understood through the inside/outside distinction. However, Heidegger shows no interest in discovering whether and what our fellow animals "really feel or think" about some matter, and whether they too might be able to transpose. Indeed, he does not even seem to be open to this as a possibility. His approach is anthropocentric and pragmatic—how can *we* know what they are *doing*. Ethological study provides access to this understanding, although it is not yet clear how.

As for the second question—"Can we transpose ourselves into a stone?"—Heidegger says that scientists and others within a naturalistic tradition do not seriously consider this question because the stone, unlike our fellow animal, does not offer us ways of going along with.[10] Here, Heidegger's relevant distinction is between the *properties* of the stone (e.g., warming or cooling, conducting electricity or resisting, breaking or melting, odor emitting or not), for which going along with makes no sense, and the *behavior* of our fellow animals (e.g., seeing, hearing, seizing, evading, building), which we can go along with in principle. Later in Chapter Four, Heidegger undercuts this distinction, saying that it is made only for human reference to the world, and is not in the world. Having a world then becomes the sole possession of the human being. Does this mean that he equates the stone and our fellow animals? He does not seriously explore the stone/animal connection, largely because his principal focus is on human *Dasein*, but also perhaps because he wants to avoid any hint of pantheism, even one that still excludes the human (*FCM*: 354).

Heidegger says that the third question—"Can we transpose ourselves into another human being?"—is not really questionable to us at all because the possibility of transposition to another human being is part of our essence and our factical existence. In *Being and Time*, Heidegger expresses this thought by saying that humans are "being-with." Biologists might wonder whether this is largely the same as saying that we are a social species, but we should note that being-with is an existential (ontological) feature of *Dasein*, while talk of "social species" is strictly ontical. As Heidegger explains, the essence of being-with can be expressed in differing ways: "going apart from one another," "going against one another," "going alongside one another." Heidegger suggests that these differing ways can lead to the illusion that there is a gap separating humans, but ontologically speaking, we are necessarily all *Dasein*—we are not autonomous individuals—and there can be no gap. But why should this apply only to humans, other than that this is what Heidegger claims? Why should it not also be, for example, an ontological feature of at least some other social species of animals? By the end of our consideration of *FCM*, Heidegger's response to this question will be clearer.

Heidegger's discussion of transposition into other humans as "part" of our essence does not include a discussion of the genesis of this transpositional "capability." But we might still ask, when and how did our ancestors come to have such an essence? In response, Heidegger might say that, unlike reception of light, transposition is *not* a capability; rather, it is of *Dasein*'s essence. In other words, there can be no human without it, so there is no sense in which a proto-human could acquire such a capability. Here again, we have a clear understanding that Heidegger is not talking about the biological human and its ancestors (human and otherwise) but about human *Dasein*, which is ancestor-less from a metaphysical (as opposed to biological) perspective. In other words, according to Heidegger, the beings that we are considering—stone, animal, human—are *not* given to us in the same manner, thereby allowing us to speculate on their interrelatedness, as science does when it assumes that everything is present-at-hand and that all living beings (at least) are connected through evolution. For Heidegger, the stone in its essence is inaccessible, the animal is accessible in a certain way (yet to be clarified), but other humans are accessible to each other.

Talk about human transference to other beings might lead one to conclude that the matter Heidegger has been considering is epistemological. But, at the conclusion of §49, he is adamant that the issue is *being* itself. In a provocative, yet insightful, statement in response to Kant and Hegel, Heidegger says, "For in-consequence, ground-lessness and fundamental concealment belong to finitude" (*FCM*: 209). In other words, while transference into others is in the essence of human *Dasein*, there are limits. Unlike God, we humans are finite, but this very finiteness, according to Heidegger, enables us to have a world that is disclosed, in part through transference, as meaningful yet different.

In §50, Heidegger returns to a consideration of how humans have access to our fellow animals. This he does through his analysis of our relationship to the dog.

> This being-with is not an existing-with, because a dog does not exist but merely lives. Through this being with animals we enable them to move within our world. … Yet when we consider the dog itself—does it comport itself toward the table as table, toward the stairs as stairs? … It eats with us—and yet, it does not really 'eat'. Nevertheless, it is with us! A *going along with*, a transposedness, and yet not. (*FCM*: 210)

Here we see that Heidegger is generating further conceptual barriers between humans and the domestic dog: one exists, the other merely lives; one allows, the other is enabled by; one comports, the other is unable to see something as what it is; one eats, the other does not really eat. Given these distinctions, it is little wonder that Heidegger says such transference from the perspective of our fellow animals cannot be a going-along-with. Nevertheless, Heidegger claims that our fellow animals, unlike the stone, do have a sphere of potential transposability, even if there is no going-along-with in that sphere. This having and yet not having of our fellow animals is what Heidegger calls "poor in world," a fundamental deprivation of world. But animals aside, are we yet clear as to what transposability really means even for *Dasein*? Toward this aim, we must better understand the essence of the human by pursuing his/her world-forming ability.

But before returning to human *Dasein* and its comparison with our fellow animals, Heidegger needs a better justification for his characterization of the being of our fellow animals as poor in world. So, he turns his ontological gaze to biology using two approaches. One is the reevaluation of ordinary biological terms—including: organ, organism, life—and their implications for understanding animality. The other is the reevaluation of ethological studies, but now viewed ontologically, whatever this may mean. Heidegger's aim is to undercut the way in which biologists typically understand these, thereby leading to his claim that they misunderstand the being of animals.

Heidegger begins §51a (p. 212) by reasserting his holism.

> The living being is always the organism. Its organismic character is what determines the unity of this particular living being in each case. The unit of life is not the cell.[11] The multicellular living being is not, as has been suggested, a community of cells. On the contrary, … they have a specific essential wholeness by virtue of the fact that they are organisms.

Heidegger's point that living beings have an "essential wholeness" seems reasonable. His concern is that some (but not all) scientists argue that this wholeness is not essential to understanding their being, that is, that reductionism can be ontologically justified. I am not going to be able to resolve this issue, but I will say that holism considered more strictly in Heidegger's frame remains a viable and reasonable approach to understanding being, so let us stay with Heidegger.

Heidegger begins by asking, "Yet, *precisely what is an organism?*" and what is its relationship to the thesis, "the animal is poor in world" (*FCM*: 212–3)? Parroting commonsensical thinking, Heidegger suggests that an organism is that which has organs. These organs might be thought of as instruments, and the organism as a complex of instruments, say, a machine. But, quickly countering this he says that human-made objects (e.g., instruments, machines, equipment) must be derived on the basis of *Dasein*'s world-formation, so he argues that it is questionable whether these concepts can be useful as metaphors for understanding the essence of animality, and if they are not useful ontologically, then they are likely to misguide the ordinary (ontical) tasks of biologists. Biologists who describe the organism as machine and their organs as instruments—he cites Wilhelm Roux and Jakob von Uexküll—suffer from "verbal confusion" that has consequences. And, further responding to the counterargument that particular scientific results make sense, Heidegger says,

> Even if they [biologists] are guided by *unclarified* and *inadequate* theories, all factual investigations of this kind will always and inevitably produce something. Results are results. Certainly, if we are simply concerned with getting a successful result—as is often the case today—then science has performed its task. But the question at issue is this: what do these results imply for our understanding of the relevant subject-area as such, and what do they contribute to our knowledge of the elementary simplicity of the *substantive essence* of animals, plants, and matter?
>
> (*FCM*: 217)

Here, Heidegger argues that the essence of the living being cannot be captured by a mechanical model.

In §52, Heidegger reasserts his holistic thinking. He proposes that in a certain sense organs have capacities, but they have them only as part of the organism. More accurately said, it is the organism that has the capacity. For example, it is the organism that sees, and not its eyes alone.[12] The capacity for seeing already includes having the organs we call eyes. The eye by itself is not a functional unit. He complexifies his argument by saying that the human capacity for seeing cannot be the same as that of our fellow animals, despite the anatomical similarities of their eyes. This is because seeing is carried out by the organism, and as there is an abyss separating the human from our fellow animals, their "seeing" cannot be equated. Of course, no biologist would claim that what the human sees must be the same as what, for example, the dog sees. But Heidegger's critique is more severe. He is saying that the dog does not see in any ordinary, *human* understanding of that term, regardless of the anatomical similarities found in the human and dog eyes.

By contrast to an organ such as the eye, Heidegger says that equipment (e.g., a pen) has no capacity-for (writing) but does possess a particular readiness-for (writing) that is made such by *Dasein* itself. The pen by itself is still a pen, but an eye by itself has neither a capacity nor a readiness. Both the organism and equipment have their essences in what Heidegger calls "serviceability," but their ways of being what they are differ, for example, "being capable" for the organism and "being ready" for equipment. However, being capable does not entail that the organism is able to produce its own organs, let alone itself, for this is just another way of avoiding the problem through vitalism. Neither mechanism nor vitalism can explain an organ or an organism. We first need a better understanding of their essences.

Near the beginning of §53, Heidegger considers the organ. He claims that biological observations support his idea that the organism, with its capacities, is prior to its organs. He mentions unicellular eukaryotes that can digest by engulfing food particles through what he calls a temporary mouth, followed (sequentially in time as one follows "eating") by a temporary stomach, a temporary intestine, and finally a temporary anal tract. He argues that these unicells are "structureless and formless" both before the "organs" arise and after they disappear. Heidegger chooses to characterize the unicell as "structureless and formless"—and he almost certainly means ontically, for what meaning would that have ontologically? But, even using the light microscopy and tissue-staining techniques of his time and earlier, it was clear that unicellular organisms are not without fine structure (E.B. Wilson 1896), and with the later advent of electron microscopy in the 1940s, Heidegger's claim of "structureless and formless" is shown to be completely misguided. While Heidegger is correct in saying that these vacuoles and organelles (to use modern terms for Heidegger's "temporary organs") do become invisible (in biological terms, they [partially] disassemble, but not all organelles do, e.g., the mitochondrion), he could have alternatively claimed that an organellar scaffolding remains intact, waiting for the whole organelle to be reassembled and reactivated on quick notice. Rather than insisting on the presence/absence of these organelles, Heidegger could also have distinguished between their temporal restriction within unicells (something he acknowledges) and their spatial localization within multicellular animals (something he does not discuss), but doing so would have cast doubt on the fundamental meaningfulness of his presence/absence hypothesis of organs. We might also wonder whether an animal's stomach is really a stomach when there is no digestion, parallel to Heidegger's claim (see above) that the eye by itself is not an eye at all, so again presence/absence may not be the right criterion. All this probably does not occur to Heidegger because he is not a biologist, but also—and here I am making a separate point—because he disavows evolutionary thinking, a claim I discuss in Chapter 3.

Despite these various confusions, Heidegger's central point is simple: "organs are precisely not like things that are fabricated and made ready. For this reason they can never be set out or set aside somewhere" (*FCM*: 225). They are not present-at-hand; rather, they can only be understood through the way of being called *life*. Organs are subservient to their various capacities for and are not, like equipment, simply serviceable and in that way complete in its being. Heidegger says, "Capacity diverts itself into its own wherefore and does so in advance with respect to itself" (*FCM*: 226). The organ has an "urge" toward functioning—that is, its subservient way of being—while equipment, although ready to function, has no such urge. Having this urge, however, is not invoking a vitalism or some form of reductionism. The urge of the organ is only such in its subservience to the organism, which, in §54, Heidegger also describes as "self-driving," "self-regulating," and "driving forces," in contrast to the explicit prescription of a ready-made piece of equipment.

While these drives can be abstracted as simply mechanical, Heidegger's approach aims to exclude this manner of being for the organism.

In §55, Heidegger complexifies his account of an organ's capacity by saying that it can only be understood within the context of the organism's essence; that is, understanding the organism's essence is prior to understanding that of the organ. For example, what a bee "sees" is not determined by its eyes alone or by the bee alone but by the bee as an organism in its environment to which the eyes are subservient. Initially without comment, Heidegger here uses a word—"environment"—that later is used in contrast to the word "world" of human *Dasein*. Seemingly harmless in its introduction, this word leads Heidegger to ask what an environment means in general for an animal, only later saying that such a question is misguided when directed to the human in its essence. It is this difference between environment and world that Heidegger later argues provides a key to understanding the thesis, "the animal is poor in world." But we also need to ask why it is that the animal has an environment while the human has a world.

In §56, Heidegger returns to the self-driving and self-regulating capacity of subservience (also considered as "self-preservation") by asking what is meant by "self"? He says,

> the capacity does not leave itself behind, it does not escape itself as it were. On the contrary: in this instinctual "toward", the capacity as such becomes and remains proper to itself—and does so without any so-called self-consciousness or any reflection at all, without any relating back to itself. That is why we say that on account of this essential being proper to itself the capacity is *properly peculiar*. We shall reserve the expression "self" and selfhood to characterize the specifically human peculiarity, its particular way of being proper to itself. … The way and manner in which the animal is proper to itself is not that of personality, not reflection or consciousness, but simply its proper being. *Proper peculiarity* is a fundamental character of every capacity. … Proper peculiarity is not an isolated or particular property but rather a specific manner of being, namely a way of being proper to oneself.
>
> (*FCM*: 233)

So, Heidegger sets up another conceptual dichotomy that separates our fellow animals from humans, a *proper peculiarity* of our fellow animals that does not relate back to itself versus the unique peculiarity in human *Dasein* that does relate back to itself and for which he reserves the words "self" and "selfhood." This distinction between our fellow animals and humans is not the result of a scientific investigation because *peculiarity* refers to (a way of) being and only indirectly (through interpretation) to beings. Here, Heidegger interprets ethological findings within an ontological frame, and his point is not to discover something about the behavior of our fellow animals but rather to disclose our fellow animals *as such*, thereby making them available for subsequent discovery as they truly are, and not as biologists have previously presented them. Importantly, Heidegger is not claiming that our fellow animals are creatures of instinct in the sense that behavior can be reduced to some autonomous mechanism, as biologists might have it; rather, the behavior of our fellow animals "remains proper to itself … without relating back to itself." Heidegger demands that we think holistically when considering instinct. Still, we might doubt whether proper peculiarity and instinct cannot be connected.

As interpreted from observations of animal behavior, our fellow animals (organisms) manifest multiple "capacities for." Heidegger mentions "assimilation, growth, inheritance of acquired characteristics,[13] propulsion, struggle with its enemies" (*FCM*: 234). Likewise, the organs of the organism manifest multiple capacities, the disclosure of which is based on observations of animal physiology. For biologists, these capacities refer to present-at-hand capabilities of the organism and the organ. But for Heidegger, these capacities must first be interpreted ontologically and can be subsequently discovered only through interpretation of the organism's capability for, which is an irreducible (holistic), non-quantifiable feature of animality.

After presenting a more justifiable way of talking about organs vis-á-vis the organism, Heidegger then says that it would be incorrect to say that an organ "has" capacities; rather, it is the organism which manifests the capacities, which in turn create the organs. Organs cannot preexist without the organism having capacities; otherwise, they are not the organs they are. But the talk of capacity need have no such material anchor as organ (remember Heidegger's discussion of temporary organs in amoeba and infusoria), so "capacity" has a priority over "organ" in our understanding. But by similar reasoning, it is also incorrect to say that an organism "has" capacities, if by this one means that the capability for it anchored in a pre-existing material body; otherwise, the organism precedes its capability for. Rather, in becoming the creature that it is, the animal (organism) is "rendered capable," that is, its being "is" a potentiality to manifest its capacities. Heidegger says,

> Thus the organism is neither "a complex of instruments," nor a union of organs, nor indeed a bundle of capacities. The term "organism" therefore is no longer a name for this or that being at all, but rather designates a particular and fundamental manner of being. We can briefly characterize this specific manner of being as a proper peculiarity with the capability to create organs. … *Being capable […]* belongs to the animal's *being actual*, to the essence of life.
>
> (*FCM*: 235)

So, ontologically speaking (as Heidegger is actually doing here), "organism" is removed from the exclusive materiality assigned to it by biologists, and, instead, designates a way of being, called *proper peculiarity*.

Relatedly, animal behaviors—Heidegger mentions hunting, grasping, and seeing—are not simply sequences of isolable, present-at-hand events; rather, these necessarily occur, and must be understood, within a broader context than the animal itself (e.g., the hawk is hunting for a vole, which is hiding or fleeing; the cat is grasping a millipede, which is struggling; the bonobo is seeing another bonobo, which is copulating with a third bonobo). Heidegger calls these "self-life" behaviors, which are distinguishable from what a stone "does." But they are also distinguishable, according to Heidegger, from what a human does, which he says are "doing and acting." By contrast, the behavior of our fellow animals is "an instinctual drivenness." Unlike biologists, Heidegger restricts the word "behavior" to our fellow animals and, instead, assigns "comportment" to humans only, making clear that he now regards all behavior as instinctual in his special sense.

> Capability is instinctual, a driving forward and maintaining oneself in being driven toward that which the capacity is capable of... The behavior of the animal is not a *doing and acting*, as in human comportment, but a *driven performing*. In saying this we mean to suggest that an instinctual drivenness, as it were, characterizes all such animal performance.
>
> (*FCM*: 237)

Importantly, this "instinctual drivenness" is not mere possibility—as it might be for humans—but "actuality" of intrinsically being capable of. The important distinction is one of time—"possibility" points toward a future, while "actuality" points to the present.

Behavior is being capable, which is an intrinsic self-retention or proper peculiarity. Our fellow animals, according to Heidegger, are intrinsically absorbed in themselves, which is a prerequisite for there being behavior in the first place. By contrast, comportment is not an absorption but a standing-in-relation-to something (Heidegger 2009: 40). In case his point is still not understood, he introduces the term "captivation," and then says, "the animal behaves within an environment but never within a world" (*FCM*: 239). "Captivation" is not a state of our fellow animals but rather is intrinsic to them. In all this, Heidegger has built a web of terms that reinforce his aim to distinguish humans from our fellow animals.

In §59, Heidegger develops this animal-behavior/human-comportment distinction by interpreting ethological studies of worker bees. I delay discussion of this important section until Chapter 4, which I devote to scientific studies of bee behavior and their ontological interpretation.

Also in §59, Heidegger clarifies what he means by captivation:

> To say that captivation is the essence of animality means: The animal as such does not stand within a manifestness of beings. Neither its so-called environment nor the animal itself are manifest as beings. The animal in principle does not possess the possibility of attending either to the being that it itself is or to beings other than itself, because the animal is directed in its manifold instinctual activities on the basis of its captivation and of the totality of its capacities. Because of this driven directedness the animal finds itself suspended, as it were, between itself and its environment, even though neither the one nor the other is experienced as being.
>
> (*FCM*: 248)

Here, Heidegger says that our fellow animals in their behavior are attending neither to themselves nor to any other beings, but rather they are directed to their own instinctual drives as their proper peculiarity. But still, Heidegger says that our fellow animals are *open* to their environment. What specifically are they open to?

Heidegger says that openness occurs in the *relationship* among the multiple instinctual drives (e.g., hunger and thirst, mating, fleeing from danger). These drives give rise to a "reciprocal drivenness" (e.g., in fleeing from danger the animal's thirst drive is inhibited), which he metaphorically describes as being encircled by a ring. This ring constitutes the total drives of the animal. But the openness of our fellow animals, according to Heidegger, is never an *attending to*; quite the contrary, it has an *eliminative* character (a *not* attending to).

In §60, Heidegger returns to the question of what it is that our fellow animals are open to. The drive of the animal, Heidegger says, is intrinsically determined by its "being driven," where this being driven is of the essence of the organism, and not to be thought of as an optional embodied trait. As drives are intrinsic to the organism as a whole, individual drives cannot be considered in isolation from other drives. Heidegger says, "The animal is *encircled* by this ring constituted by the reciprocal drivenness of its drives" (*FCM*: 249). That our fellow animals are "unable to attend to" (i.e., unable to relate to the other as it is what it is) is to be understood more fundamentally as "eliminative"; that is, the other appears, but cannot be seen, as what it really is, and thus it is eliminated. For example, Heidegger notes (*FCM*: 250) that, in some cases, a female insect devours the male after insemination. According to Heidegger, this indicates that the female insect cannot relate to the male "simply as a living creature" but only as a sexual partner or prey, where here Heidegger uses "sexual partner" and "prey" as stand-ins for that which is not manifest to the female as such but as that which disinhibits the female insect to evince a certain response. We could also say that the female insect does not see the male as it is what it is, but only its utility. Here, we might question whether it is reasonable for Heidegger to generalize from the intense situation of mating to an insect's entire life, but for Heidegger, the insect's intrinsic behavior is always an elimination, unlike the comportment of humans, which can be positive in the sense of relating to something as what it is. So, we return to the question of what the female insect, for example, is open to? And relatedly, "What is this encircling [ring] like, if a relation to other things is not merely sustained, but constantly brought about by this encircling?" But Heidegger also says, "Eliminative behavior thus represents the continual production of an emptiness" (*FCM*: 252). Our fellow animal "does announce itself as something that relates to other things, and does so in such a way that it is *somehow affected* by these other things" (*FCM*: 253). Yet, it does not comport to other animals as such. Other animals are never manifest as the beings they are. Rather, an emptiness is produced. What is going on here?

A meaningful ontological interpretation of behavior, such as for the case just mentioned, requires that ontical observations be both accurate and sufficiently complete so as not to be misleading. For example, failure to note that a male spider is not always devoured by the female is misleading in that it makes the behavior seem undeniably instinctual in its inevitability (Zhang et al. 2022). But even so, interpretations of behavior/comportment can still remain underdetermined by the observations. As an example, when a husband puts himself in harm's way for the benefit of his spouse, the behavior/comportment itself can be interpreted as an identification with the other self as such (a "freedom"), a blind concern for perpetuating one's own genes (a "captivation"), or both in combination. Might there be bias in which interpretation is preferred, depending on whether similar observations apply to humans or to insects? Heidegger's use of ethological findings clearly demonstrates that his ontological interpretations are dependent on particular ontical understandings about what is "really" going on.

Heidegger's proposal of an encircling ring for all of our fellow animals is important because it puts in ontological terms what many people already assume about our fellow animals, even though few articulate this as clearly as Heidegger. What they assume, like Heidegger, is that there is an unbridgeable chasm separating the being of our fellow animals from that of humans, and that there is no way to challenge this. Inside this ring of darkness, Heidegger grants dynamism, but without awareness to our fellow animals.

According to Heidegger, the animal, although self-absorbed, is not "encapsulated"; rather, it is still open or "capable for" in the limited sense that certain (external) stimuli are able to disinhibit, thereby leading to behavior. These stimuli are those, and only those, that affect the animal's instinctual drives, of which there are many. Together, they constitute the animal's encircling ring. Furthermore, the disinhibition of one instinctual drive may effect a response in another, giving rise to an apparently complex behavior, but still within the animal's encircling ring and not to be confused with comportment. Heidegger's point about disinhibition (in contrast to activation, a term he does not use) is that that which disinhibits behavior is not apparent (or, visible) to the animal; in his terminology, there is no "attending to." The animal does not relate to something outside itself as that something, but rather, only responds to its own disinhibiting ring. The life of the animal, then, is the struggle to maintain this ring if the animal's life is to continue. Heidegger's talk about (animal) life and its essence may appear to be deconstructive (e.g., he speaks of disinhibition, instinct, and stimuli), but in the ontological sense he is describing structural moments of the encircling ring. Such moments identify a conceptual articulation that is not based on any material or mechanistic understanding. This should be obvious because being is not a being.

Heidegger dramatically concludes §60 by saying:

> Thus we can see that the circumscribed range of possible disinhibition or, as we put it, the disinhibiting ring, is intrinsic to the animal itself; that in its factical life amongst beings in each case the animal itself carves out a quite specific encircling ring within which it can be stimulated. … We must say that life is nothing but the animal's encircling itself and struggling with its encircling ring, a ring by way of which the animal is absorbed without its ever being with itself in the proper sense.
>
> (*FCM*: 257)

With this, it would appear, the abyss (chasm) between our fellow animals and humans is complete.

In §61a, Heidegger further clarifies how his metaphysical interpretation of our fellow animals differs from that of a scientific view:

> The *organization of the organism* does not consist in its morphological, physiological configuration, in the formation and regulation of its forces, but first and foremost precisely in the *fundamental capability for self-encirclement*, and thus for a quite specific openness for a circumscribed range of possible disinhibition. … This capability for encirclement is the fundamental characteristic of the animal's actual being in every moment of its life-span. But if we consider the organism as the morphological unity of the [animal] body, which is what usually happens and especially today, then we have still failed to grasp the decisive structure of the organism.
>
> (*FCM*: 258)

Contra the typical reductionist presumptions of biologists, who grasp the animal as a collection of interacting components affected by a separate environment and who, thereby, fail to "see," but only presume, the animal as animal (i.e., in its essence), Heidegger says that the animal "is" its capability for self-encirclement by a disinhibiting ring (captivation). This ontological framework identifies the essence of *all* of our fellow animals and provides the proper framework for interpreting scientific investigations. In this context, Heidegger specifically criticizes Darwinism (*FCM*: 260), an issue I address in Chapter 3.

Heidegger summarizes the organism's way of being as captivation in what he calls its six structural moments:

1. A *withholding* of the manifestness of beings.
2. A *being taken* (instinct) in openness to other things.
3. An *absorption* in all of its drives.
4. An *openness* for that which disinhibits within the encircling ring of drives.
5. The *encircling ring*, circumscribing all of the animal's instinctual activities, brings forth the essential character of life itself, including its self-preservation.
6. Captivation as the possibility of all behavior.

An ontical parallelism with a biologist's understanding of instinct should be very apparent.

In §61b, Heidegger acknowledges that some biologists have partially advanced a proper understanding of the organism and of the organ. Principal among these is Jakob Johann von Uexküll (1864–1944), who today could be described as an ecologist, ethologist, and theoretical biologist. Von Uexküll shared Heidegger's holistic understanding of the natural world and considered all animals as being in an environing world (*Umwelt*), partially of their own making, within which they carry out their lives (von Uexküll 1982, 2010, 2018). Heidegger agrees with some of von Uexküll's ontical understandings of the natural world, although not his musings on the "inner life" of our fellow animals or his understanding of humans as animals. According to Heidegger, von Uexküll still fails to understand the animal in its essence. Hence, von Uexküll's talk of an animal's "world" becomes, for Heidegger, their "environment." The world of the human becomes a special world that is specific to human *Dasein* as being-in-the-world. In my own Chapters 4 and 5, I address new ethological findings and discuss how these might impact Heidegger's understanding of the organism.

As concerns understanding the organ, the embryologist Hans Driesch (1867–1941) warrants special note by Heidegger when he shows early stage sea urchin embryos to be capable of self-regulation. In particular, these embryos compensate for the human-induced removal of a blastomere by regenerating a normal, albeit smaller, embryo with the proper number of cells. This fits nicely into Heidegger's holistic frame, where the organism *directs* cell behavior, but Driesch himself ventures toward vitalism. However, Heidegger also notes that subsequent experiments described by the embryologist Hans Spemann (1869–1941) complexify any simple interpretation. In particular, Spemann's student Hilda Mangold (1898–1924) performed experiments which illustrate the ability of the early stage (blastula) newt embryos to regulate their cells' identities, depending on their position within the embryo, thus confirming Driesch's findings. However, when performed on later stage (gastrula) embryos, the cells' identities remain fixed, impervious to their new location. Heidegger notes that the problem of the organism's organization now has "a more comprehensive and profoundly conceived basis" (*FCM*: 266) that leads our understanding "in a quite new direction" (*FCM*: 262). These experiments argue neither for nor against holism but rather indicate that, within the developing embryo, individual cell fates become fixed or determined at different life stages, depending on the species. Within an evolutionary frame, this occurs for reasons related to the species' particular life history. Heidegger's lack of further comment on Mangold and Spemann's findings suggests that he was unclear how to respond, or that they were outside his ontological purview, since holism and lack of cell-fate determination are not equivalent ideas. But contra Heidegger, many reductive biologists continue to this day to view the cell as the organism's basic unit, and Heidegger's influence on the subsequent revolution in our understanding of evolutionary cell fate determination has been nil (e.g., Davidson 1968; Peter and Davidson 2015).

In §61c, Heidegger turns to a consideration of organism as process, or more generally, as motion. In this, he emphasizes the drivenness of the instinctual drives—that is, not just the drives themselves—as also belonging to captivation. In other words, captivation should not be viewed as a rigid shell straightforwardly defined by the totality of its drives as previously considered but also an activity that waxes and wanes, depending on the organism's circumstances, for example, fleeing a predator or resting. Heidegger is adamant that in talking about motion he is not referring to the life history of the organism or its development as understood by biologists who observe or perturb other life forms. The essence of the organism can only be revealed by understanding captivation, which grounds any and all ontical understandings of natural history and development. To avoid any ongoing confusion, it ought to be reemphasized that biologists and Heidegger use the term "organism" in different ways. For reductive biologists, organisms are first-and-foremost embodied collections of spatially and functionally separable organs, tissues, cells, and their underlying mechanistic processes. For Heidegger, an organism refers to a kind of wholeness, ontologically defined by captivation and proper peculiarity, that conceptually demarcates living forms and their life activities.

Heidegger makes the further novel claim:

> The species is by no means simply a logical schema under which the actual and possible particular individuals are subsumed. The *species-character* is rather an essential *character of the being of living things*, and one that precisely finds expression in what we have come to know as the fundamental structure of animality: captivation.
>
> (*FCM*: 265)

Here, "species" refers only to our fellow animals, and not to the human species. We might ask, what is added to our previous understanding of captivation of the individual by saying that it is also essential to the species? In reply, Heidegger says, "the species as such is thereby better protected and better equipped in relation to its environment" (*FCM*: 266). In saying this, Heidegger is indicating that animal species, like their individual members, do have a relation to an environment, but this relation is—in making an analogy with individual animals—"instinctual." By contrast, human *Dasein*—and only human *Dasein*—can have a history (*Geschichte*). But to have a history, human *Dasein*—in its authentic form—is always "in each case mine" and never just an organism, let alone simply a member of a species. Even in its inauthentic form, human *Dasein* has a "herd"-self. On the other hand, as all animals are captivated, it makes little difference whether we refer to them as individuals or as species.

Heidegger concludes §61 by returning to a distinction he has already introduced in *Being and Time*: between "death" (also translated as "ontological death") as a way of life for *Dasein* and "perishing" as a biological process applicable to our fellow animals but not to humans. (He reserves "demise" as the biological end of human life.) For *Dasein*, there is a motility, not in the sense of physical motion but in the sense of progression toward ontological death. Apparently, there can be no such motility—ontologically speaking—for our fellow animals. But what about the other end of life's spectrum? Is there an ontological birth for *Dasein*? This question is not addressed in *FCM*, but elsewhere around the same time, Heidegger says that even a child can be "primitively understood" as in *Dasein* (Heidegger 1988), although lacking "a specific luminosity" (Heidegger 2024: 87) or in a "twilight" condition (Hatab 2014). Even so, Heidegger is cautious in his interpretation of childhood *Dasein*. He says (Heidegger 2024: 88-9),

> It is not the case that the child first emerges from an enclosed subject in the course of its first weeks and arrives at objects; rather it is already—and not just when it has been torn from its state of latency—directed out toward …; it is

> already outside alongside .... A being of some kind is already manifest to the child, although there is as yet no comportment toward this being, no turning toward it ensues. Turning away and warding off and this self-centered need for restfulness, warmth, and sleep have a quite peculiar negative character. Until these phenomena such as warding off, turning away, and resistance have been clarified in their ontological structure, we cannot begin to interpret a state such as that of the child in its essence.

For Heidegger, phrases like "directed out toward," "outside alongside," "turning away," "warding off," and "resistance" are not scientific terms; they are closer to adult *Dasein*'s existentialia (see List 1.1). While Heidegger thus acknowledges temporal changes in human *Dasein*, I would suggest that by Heidegger's thinking the origin of *Dasein* necessarily remains outside the scope of scientific investigation—both of ontogeny and phylogeny—and, more to the point, a misguided venture given the ontological abyss between *Dasein* and non-*Dasein*.

Heidegger begins Chapter Five (§§62, 63) by summarizing the dilemma presented in the previous chapter: both humans and our fellow animals, but not the stone, have accessibility to something other than themselves. But still, do our fellow animals have a world or not? What is the meaning of "world"? What does Heidegger's second thesis really mean? Heidegger resolves this dilemma by saying that while our fellow animals indeed have access to ("are open to") something outside themselves, they (unlike human beings) are unable to experience ("are not open to") it as the manifestation of other beings. Rather, the access of our fellow animals is expressed through their instinctually driven behavior that occurs within an openness to a disinhibiting ring. What they have is not a world but rather a being open to disinhibition by that which disinhibits, without experiencing that which disinhibits as such. The essence of our fellow animals is being open in captivation. Our fellow animals do not have "less" of a world than we; rather they do not have a world at all. The thesis that "the animal is poor in world" actually overstates the capability of our fellow animals. It is human beings who can have less of a world in a quantitative sense, but only because they can have a world in the first place. The seeming dilemma arises only because we humans imagine that our fellow animals are like ourselves. So, Heidegger says that his understanding of the essence of our fellow animals is

> a merely comparative definition of animality in terms of poverty in world, insofar as the animal is viewed from the perspective of man to whom world-formation belongs. Our thesis that *the animal is poor in world* is accordingly far from being a, let alone the, fundamental metaphysical principle of the essence of animality.
>
> (*FCM*: 271)

But if captivation is "a merely comparative definition of animality," then what is an essential understanding? Heidegger says,

> We must *leave open the possibility* that the proper and explicit metaphysical understanding of the essence of world compels us to understand the animal's not-having of world as a *deprivation after all*, and to discover *poverty* in the animal's specific manner of being as such.
>
> (*FCM*: 272)

But almost immediately thereafter, Heidegger appears to argue against this, finding inspiration in the writings of St. Paul, to which Heidegger responds:

> For with the animal's being open for that which disinhibits, the animal in its captivation finds itself essentially exposed to something other than itself, something that can indeed never be manifest to the animal either as a being or as a non-being. Rather that which disinhibits, with all the various forms of disinhibition it entails, brings an *essential disruption* into the essence of the animal.
>
> (*FCM*: 273)

In other words, Heidegger repeats his claims about withholding the manifestness of beings in animal captivation, but then he adds something new. He says that the disinhibitor effectuates "an essential disruption into the essence of the animal"; that is, the disinhibitor disrupts the animal's very being. Do our fellow animals now have two structural moments, not just one—(1) withholding manifestness and (2) the disruption of being? Could the disruption be thought of as an intrinsic openness to manifestness of a reduced sort? Unfortunately, Heidegger does not develop his thinking about disruption, but it does lead him to say that withholding manifestness in itself can no longer be thought as the essential ground of the whole (i.e., it must be thought together with the disruption), and, therefore:

> We can now reply that in the last analysis we have not yet clarified the essential organization of the organism sufficiently at all, so as to be able to decide the significance of this withholding, and that we cannot clarify it until and unless we also take into account the fundamental phenomenon of the life process and thus death as well.
>
> (*FCM*: 273)

Heidegger's claim appears to be that the essence of organism remains unclear because of a lack of clarity as regards "the significance of this withholding [of manifestness]" for the animal itself in light of the newly discovered disruption. But what remains the case for Heidegger is that no world is manifest to our fellow animals, even if that is but one moment of an animal's way of being.

Chapter Six (§§64–76) of PART TWO is relatively large—it occupies 25 percent of the book—and is the culmination of Heidegger's central aim to make available to his students and readers the possibility of transforming their own inauthenticity into *Dasein*'s authenticity, now viewed philosophically (as introduced in PRELIMINARY APPRAISAL) within the frames of profound boredom (as explained in PART ONE) and the animal's lack of world-formation (as presented in Chapters Three–Five of PART TWO). In this final chapter, Heidegger explains that and how humans are world-forming, whether in their authenticity or inauthenticity. Despite the importance of this chapter to Heidegger's project vis-á-vis human *Dasein*, my commentary emphasizes his mention of our fellow animals rather than his deep dive into Ancient Greek philosophy. However, to the extent that I am claiming (as I will!) that our fellow animals too are in *Dasein*, these other parts of this chapter remain relevant.

In §64, Heidegger returns to the relationship between manifesting a world and seeing "something *as* something." He says both of these are "fundamentally closed to the animal" (*FCM*: 274). Without access to using the "as," our fellow animals—unlike humans—are unable to attend to whatever they encounter (i.e., to whatever it is that disinhibits their encircling ring) in the sense of "letting it be" or "not letting it be."

In §65, Heidegger turns to *Dasein* in its everydayness. He says that humans ordinarily experience their world as undifferentiated in their contemporary boredom (see Table 1.1). Heidegger's point of discussing animality earlier in PART TWO (§§45–63) is to emphasize that our fellow animals are fundamentally different from humans, something that *Dasein* in its everydayness (i.e., in its contemporary boredom) fails to grasp. By living in an undifferentiated world, Heidegger says, contemporary *Dasein* "exists," but only analogically, in the world as animals.

In §66, Heidegger says that each kind of animal has its own encircling ring. Yet, these different rings overlap with each other, thereby giving rise to an abundance of contents and relations. These contents are not to be equated with the reductive descriptions of biologists, nor are the relations to be directly equated with inter-species or species/environmental interactions as described by ecologists, since Heidegger's analysis is ontological, not ontical. It is important to reemphasize that Heidegger has already claimed there to be a *single* essence of our fellow animals, namely, openness to captivation, so the multiplicity of encircling rings should be thought of as different aspects of an underlying unity. Heidegger says that human *Dasein* intrinsically has a "peculiar transposedness" which allows it to enter the "contextual ring of living beings" without itself "having" such a ring. The error of everyday *Dasein* is to suppose in a confused manner that humans too have such an encircling ring. Here Heidegger is not making an argument, but rather clarifying that the inter-species differences which interest him are not those that interest biologists, and that his interests are primordial to those of the biologists. Of course, we might still question how Heidegger came to the ontological claims that he did.

In §67, Heidegger returns to the question of world, distinguishing its accessibility as a capability intrinsic to *Dasein* from a flawed naturalistic conception of world, which is grounded independently of *Dasein* in the manifested beings themselves. In other words, according to Heidegger, beings are part of the world only insofar as they are accessible and open to *Dasein*. By contrast, our fellow animals are denied accessibility and openness to beings and world. Understanding accessibility to entities requires a holistic approach, in which *Dasein* and other entities *together* are manifest in their unity, a unity that is an aspect of the larger unity he calls "world." Heidegger's unusual choice of the word "manifestness" indicates that world, as presented in *FCM* but not elsewhere (e.g., *Being and Time*), is neither an abiding entity (or collection of entities) nor a dynamic happening (i.e., process), as in "to manifest something"; rather, manifestness is the result of manifesting, of giving rise to presencing or being of entities.

At the end of §67, Heidegger again notes that even inauthentic human *Dasein* has a world, albeit uniformly depauperate, and that there is always the possibility, if one is suitably prepared, for inauthentic humans to be drawn into a moment of vision (Table 1.1), in which a plurality of beings become manifest in their being, including one's own self.

In §68, Heidegger finalizes—but only for this period in his intellectual life-journey—his formal indication of "world" by emphasizing that world in its manifestness has two structural moments.

> World is not the totality of beings, is not the accessibility of beings as such, not the manifestness of beings as such that lies at the basis of this accessibility—world is rather *the manifestness of beings as such as a whole.*
>
> (*FCM*: 285)

In this updated understanding, Heidegger adds the "as a whole" to indicate that world always has a "characteristic wholeness." But what does this mean? In a preliminary way, Heidegger says that it relates to "*Dasein* in man is world-forming" in three senses: (1) *Dasein* brings forth world; (2) *Dasein* gives a particular view of world; and (3) *Dasein* constitutes and engages with world. In overly simple terms, we can say that the "as a whole" provides the background out of which world-formation of individual *Dasein*s and other beings can occur. It does not mean that *Dasein* sees the whole, whatever that God-centric idea may mean, but rather that *Dasein*'s view is presented as *unified.*

Section 69 returns to a more in-depth discussion of the "as," which Heidegger has already indicated to be unavailable to the animal. He notes that this "as" points to formal indication, not to intentionality as typically understood. Formal indications, in turn, always point to transformations in *Dasein*. Therefore, as a listener or reader of Heidegger, it is *my Dasein* that must be challenged, not some *Dasein* in general. Taking a formal indication to be general—one that is not anchored in a particular *Dasein*, as scientific concepts are ordinarily interpreted—leads to addressing groundless questions, e.g., Heidegger cites "human freedom" as one full of irresolvable conundrums (*FCM*: 296). Heidegger's central conclusion about the use of formal indication in *FCM* is that we will never be able to address the phenomenon of world (or indeed of individuation or finitude) directly. World is simply not some being that is present at hand, nor is it intrinsically a part of *Dasein*, although *Dasein* is necessarily linked to world.

If world and *Dasein* are not present at hand, then what are they? Heidegger says that they are not isolated concepts nor is there an immanent connection between the concepts. Rather,

> *The one and only originary interconnection of concepts* is already *established through Dasein itself.* The vitality of this interconnection depends upon the extent to which Dasein in each case comes to itself. … The intrinsic conceptual interconnection is that of the history of Dasein itself, something which, as history, transforms itself. This is why formally indicative concepts and especially fundamental concepts can in an exemplary sense *never* be taken *in isolation.*
> (*FCM*: 298)

As an example of a seemingly unexpected source of conceptual interconnection, Heidegger turns to the role of the "nothing" in understanding *Dasein* (*FCM*: 299). Here, the "nothing" is not the empty nothingness based on immanent thinking but rather an absence that actually matters to a particular *Dasein* through its interconnection with "sensory" aspects of world, for example, the absence of one's lover on a summer evening or the regret of a former choice that is no longer possible. For Heidegger, presence and absence are not opposed as they are when thinking present-at-hand. Heidegger does not talk about our fellow animals with respect to the nothing; rather, the very idea of disinhibition of an encircling ring has an all-encompassing darkness built into it, but this darkness is not the nothing of *Dasein*. The darkness is without meaning to the animal; it is a kind of permanent blindness to that which is influencing our fellow animal. But is access to such "nothing" really inaccessible to our fellow animals? Heidegger's view of our fellow animals is indeed dark, and as I later argue, unwarrantedly so.

In §71, Heidegger delves more deeply into the "as," saying that it is based on discourse, which in turn enables understanding and comportment (§72). In *Being and Time* (pp. 203–4), Heidegger says "Discourse is the Articulation of intelligibility. Therefore, it underlies both interpretation and assertion." We might wonder whether discourse outside of interpretation and assertion can occur without the "as." Christoph Demmerling (2021: 237) says, "Heidegger's conception of discourse oscillates between a broadly linguistic (pre-predicative and non-assertoric) and a pre- or non-linguistic pragmatic understanding thereof." Despite this ambiguity over the role of language for human *Dasein*, Heidegger maintains that discourse is denied to our fellow animals. Heidegger says that those who interpret the sounds of our fellow animals as (meaningful) words misunderstand animal captivation. Comportment, in contrast to behavior, is "free" in the sense that one can either be "open for" or "one can fail to be open for" something as something, just like language itself can point out by revealing something or point out by concealing something.

Implicit in what Heidegger says is that our fellow animals can be influenced by their past only as mediated through their present situation. And any hint that their behavior is linked to a future is a strictly imaginative construct by a confused human. Most fundamentally, the discursive use by humans of the "is" has no meaning for our fellow animals. As Heidegger explains Aristotle, the "is" is not simply a naming of something; rather, in its meaningful function it serves in synthesis, connectedness, and unity.

In §74, Heidegger says that world is not something that can ever be directly observed because its essence—the prevailing of world—is more originary than beings. World manifests not just those beings that are at hand but all those that matter to us in some way or another. In our everyday being, these beings are typically undifferentiated, either "invisible" or indistinguishable from other beings in their essence, but they are there. Like fundamental attunement, having a world requires a transformation of *Dasein* itself, a release from non-fundamental attunements and from misguided metaphysical directions.

In §75, Heidegger articulates his claim that being—in this case, *Dasein*'s being—is necessary for any understanding of beings and is not simply something that philosophers like himself speculate about. Heidegger argues that an exclusive focus on beings, like scientists have, leads to a diminution in world itself through its undifferentiatedness, for example, by considering humans and our fellow animals to be of the same sort. But it is not just being and beings that matter but also their *distinction*, which he addresses by making 9 overlapping points (*FCM*: 357–8). In summary form, Heidegger says,

> The ontological difference is the difference sustaining and guiding such a thing as the ontological in general, and not a particular distinction that can or must be made within the ontological.
>
> (*FCM*: 359)

What about the ontological difference for our fellow animals? Well, Heidegger has already said that our fellow animals cannot relate to beings as beings, so for them there can be no ontological difference of which they are "aware." But, according to Heidegger, our fellow animals do "relate" to beings in their openness to captivation, so we might still wonder whether there is something meaningful in the difference between the two, an ontological difference that is distinct from that of human *Dasein*. About this possibility, Heidegger is silent.

In §76, Heidegger turns to "a still more originary manner of thinking" in which he proposes that "'being transposed in Da-sein' be thought of as 'projection.'" Not to be confused with ontical thinking, "originary projection" is what "makes possible all ordinary projection and goal-setting." Projection opens up making-possible as a prefiguring of actuality but itself has no goal. In relating the possible and the actual, projection

> is the proper occurrence of that distinction between being and beings. The projection is the irruption into this "between" of the distinction. … The projection unveils the being of beings.
>
> (*FCM*: 364)

"Projection" is Heidegger's final word on human *Dasein*'s world-forming ability. The irruption toward possibilities that characterizes projection is *Dasein*'s particular way of life, now called "ek-sistence" to distinguish it from the unhyphenated form ("existence") that is burdened by previous philosophical misunderstandings, including existentialist ones. Similarly, Heidegger introduces *Da-sein*, in distinction to *Dasein*, to emphasize a more authentic ek-sistence. Through projection, humankind in its essence is unable to remain in place but is instead constantly entering new possibilities. Humankind is always in transition. Unlike our fellow animals, so Heidegger would say, humans are not locked in a present but rather for them, there is always a "having been" and a "future."

This important, final section of *FCM* does not mention our fellow animals. But still, we can ask whether projection as necessarily confined to human *Dasein*—or even more restrictedly to *Da-sein*—should be considered obviously correct? While Heidegger does distinguish between projection as originary and as thought ontically, it seems unlikely to me that we finite humans can understand originary projection except through an interpretation of determinate ontical projection (and projecting). Has Heidegger displayed adequate openness to the possibility of ontical projection in our fellow animals, and, if so, how would that be interpreted ontologically? Has Heidegger demonstrated that our fellow animals are locked in the present? In any reinterpretation, we should be fair to Heidegger's insights but not so much that we are blind to the abilities of our fellow animals.

Chapter 2

HEIDEGGER'S UNDERSTANDING OF AN ANIMAL'S WAY OF BEING IN HIS OTHER PUBLICATIONS

Heidegger's thoughts about our fellow animals as enunciated in the *FCM* lectures of 1929–30 do not appear *de novo* but rather are presaged in his earlier transcribed lectures, books, and essays. Similarly, sprinkled throughout many of Heidegger's post-*FCM* writings are comments about our fellow animals. In this chapter, we explore whether these other writings present different ways that Heidegger thought about our fellow animals from those in *FCM*. Any differences might illuminate his personal biography, intellectual history, and philosophy.

Section A. Writings Prior to The Fundamental Concepts of Metaphysics (1929–30)

Basic Problems of Phenomenology (Winter Semester 1919/20)

(Lectures of 1919–20; Heidegger 2013a)

These early lectures presage important themes that are later developed in *Being and Time* and *FCM*: (1) phenomenology as the originary approach to interpreting life, where "life" is an early term for human *Dasein*; (2) human factical life (or "facticity") as central to understanding life-world, that is, any particular world in which we find ourself; and (3) articulation of the structural moments of a life-world (i.e., self-world, with-world, environing-world), which is related to the later articulation of being-in-the-world as found in *Being and Time*. These lectures are informative of Heidegger's earlier understanding of world, which differs from that in *FCM*. In *FCM*, the task is to demonstrate that world formation is not applicable to our fellow animals, for which a formal approach to world is particularly suitable; whereas, in the lectures of 1919–20, Heidegger is developing his understanding of world only for human *Dasein*. Here, the particularities of the different possible worlds (the "life-worlds" of, for example, biology and Christianity) become central if one is going to have a meaningful conceptual understanding of world. As our fellow animals are not directly addressed, we may question whether these lectures even relate to Heidegger's thinking about our fellow animals.

Basic Concepts of Aristotelian Philosophy

(Lectures of 1924; Heidegger 2009)

Linking Aristotle's philosophy to Heidegger's own understanding of phenomenology, this book makes several interesting comments about our fellow animals. However, understanding these presents a challenge as to whether Heidegger is interpreting Aristotle in a manner that reflects his own thinking, or whether his interpretation of Aristotle's thought may—in places—be peripheral to or even in conflict with his own. Most centrally, humans are animals by Aristotle's thinking; but what about Heidegger in 1924?

Heidegger (2009: 14, 40) says,

> A living thing is not simply at hand (*vorhanden*), but is in a world in that it has its world. An animal is not simply moving down the road, pushed along by some mechanism. It is in the world in the sense of having it. *The being-in-the-world of the human being is determined in its ground through speaking.*

And,

> The relatedness of animals to the world is precisely that which brings animals in their being genuinely into being-there.

Here, Heidegger is clearly saying that our fellow animals are both in their world and being-there. But since these are ontical claims, we might still wonder how they inform his later claim of an ontological abyss. He also says (based on Aristotle) that "the human being is determined in its ground through speaking" (later referred to as "saying" and "language") in contradistinction to our fellow animals, who use only vocal announcing (*phoné*) or broadcasting. Heidegger says that our fellow animals may even respond to these broadcasts and, by so doing, have a definite orientation (p. 68), but such responses remain instinctual. "Saying" or "speaking" later becomes for Heidegger a fundamental existentiale of human *Dasein* (only), and not simply a character of one particular animal species, as for Aristotle. In this essay, Heidegger is developing a vocabulary that will later explicitly support his claim of an ontological abyss.

"Wilhelm Dilthey's Research and the Struggle for a Historical Worldview"

(Lectures of 1925; Heidegger 2002c)

In these lectures, Heidegger (2002c: 163) says,

> Every living creature has its environing world not as something extant next to it but as something that is there for it as disclosed, uncovered. For a primitive animal, the world can be very simple. But life and its world are never two things side by side; rather, life "has" its world. Even in biology this kind of knowledge is slowly beginning to make headway. People are now reflecting on the fundamental

> structure of the animal. But we miss the essential thing here if we don't see that the animal has a world.

Central to understanding this quotation is that it is an *ontical* interpretation that reflects Heidegger's understanding of what the biologist von Uexküll is saying, namely, that both humans and our fellow animals have their environing worlds. As such, all life is to be seen first and foremost as embedded in a web of relations, contra the reductionist, mechanistic thinking of some other biologists, such as Wilhelm Roux. While Heidegger agrees that the environing worlds of different species are nonidentical, when considered ontologically (i.e., when considered within the domain of philosophy), as he does later in *FCM*, he finds that there is an abyss specifically separating the world of humans from the environmental variations of *all* of our fellow animals. And so, he eventually clarifies his own thought by no longer applying the word "world" to our fellow animals, either ontically or ontologically.

Prolegomena to the History of the Concept of Time

(Lectures of 1925; Heidegger 1992a)

These lectures provide a few examples that reinforce what I have already said. For one,

> We can say that the snail at times crawls out of its shell and at the same time keeps it on hand; it stretches itself out to something, to food, to some things which it finds on the ground. Does the snail thereby first enter into a relationship of being with the world? Not at all! Its act of crawling out is but a local modification of its already-being-in-the-world.
>
> (Heidegger 1992a:166)

The key point is that the snail's (environing-)world is never something separate from the snail itself, but is always being-in-the-world. But by the time of *FCM*, such a holistic, ontical interpretation is transformed into an ontological statement—the words "relationship," "world," and "being-in-the-world" are confined to human *Dasein* and, as concerns our fellow animals, they live in darkness amid the intersection of encircling rings. And ontical similarities between the snail and the human become nonessential when considered ontologically, thereby reinforcing Heidegger's distinction between humans and our fellow animals.

Being and Time

(First published: 1927; Heidegger 1962)

Being and Time is a book about the question of the meaning of being vis-á-vis human *Dasein*, with very little directly said about our fellow animals. However, as brief as its comments are in this regard, what is said provides a prelude to what appears in *FCM*. In *Being and Time*, Heidegger says that human *Dasein* discloses—with the "help" of being—its own being-in-the-world, and thereby is able to discover entities in that world. Meaning or meaninglessness accrues to this disclosure or its absence, but only for *Dasein*. For, "all entities whose kind of Being is of a character other than Dasein's must be conceived as *unmeaning*, essentially devoid of any meaning at all" (p. 193). While our fellow animals are not mentioned among Heidegger's examples of non-*Dasein* entities, their inclusion is implicit.

Later in *Being and Time* (p. 396), Heidegger mentions the challenges of understanding ontologically the way in which the senses (e.g., hearing, touching) can operate "in something that merely has life" and how and where the being of our fellow animals can be constituted "by some kind of 'time.'" In *FCM* (p. 237), Heidegger will more boldly claim that the senses of our fellow animals reflect mere behavior, contra their central role in the comportment of human *Dasein*. Likewise, discussion of time (ontologically conceived as temporality; Blattner 2005), and its central role in being, will be restricted to *Dasein*, leaving time as an unsolved problem for our fellow animals.

The Basic Problems of Phenomenology

(Lectures of 1927; Heidegger 1988)

Here, Heidegger (1988: 190) makes interesting comments about our fellow animals in distinguishing them from human *Dasein*. He says,

> the animals can be affected on seeing the human image in the mirror just as though they had caught sight of the man himself and therefore can fear him or fawn upon him with gestures, nevertheless they do not apprehend what is thus given *as* true or false but solely as similar, and in this they are not subject to error.

Heidegger is saying that our fellow animals can be deceived about the image in the mirror, as humans can be as well, but that our fellow animals can never know that they are incorrect. Given that Heidegger is here describing a scientific observation, it remains open to being falsified, as indeed it has been for some species of our fellow animals (see my Chapter 5).

"On the Essence of Ground"

(Essay of 1929; Heidegger 1998c)

In this essay, Heidegger (1998c: 126) denominalizes world: "*Freedom alone can let a world prevail and let it world for Dasein*. World never *is*, but *worlds*."

World in its dynamism now takes priority over thinking world as enduring. But this dynamism is not ontical; rather, it is ontological. A historical comment may help clarify this. In *Being and Time* (p. 93), Heidegger's plural usages of "world" are on both sides of the ontological difference. A few years later, in *FCM*, world as "the manifestness of beings as such as a whole" is interpreted by Heidegger as being ontological only. However, "manifestness" as being the result of something assumes that entities already have some sort of intelligibility to which *Dasein* becomes open, and we could wonder how this could be known without prioritizing the world's enduring aspects. In "On the essence of ground," Heidegger arrives at a different understanding, not one where world goes from one state to another, but rather the very emerging itself ("is-ing"; Withey 2017). The "world 'worlds,'" then, refers not to manifestness of entities as in *FCM* but rather to the very projecting of entities onto meaning. In so doing, Heidegger overcomes the conflict between a preexisting and enduring world (e.g., think of the commonplace naturalistic understanding) and the world as ontological dynamism. Of course, understanding this dynamism is restricted to the realm of human *Dasein*. But is this really the case, and how could one know without some recognition of the world as enduring (e.g., as studied by ethologists)?

Section B. Writings After the Fundamental Concepts of Metaphysics

As frequently noted, Heidegger's thinking changes after *Being and Time*, focusing less on *Dasein*'s being and its world and more on being as interpreted through *Dasein*. *FCM* (see my Chapter 1) and "On the Essence of Ground" (see Section A above) already illustrate movement beyond *Being and Time*'s conceptual ambiguity about world, although human *Dasein*'s being remains central in both texts. My interest in Heidegger's post-*FCM* works remains focused on our fellow animals, and not on Heidegger scholarship generally. Nevertheless, a discerning understanding of the later-Heidegger's view(s) of our fellow animals would be depauperate without some broader context. As preparatory for understanding his individual works, I present an interpretative list of significant changes in Heidegger's word and phrase usage from the time of *Being and Time* onward (see Table 2.1 and the references therein). Some changes augur a potentially different understanding of our fellow animals, and with respect to humans these changes might suggest either more or less affiliation between human *Dasein* and our fellow animals. I emphasize, however, that these suggestions are not Heidegger's but rather have been gleaned by myself and others. In what follows, I make these suggestions upfront, and then indicate the nature of Heidegger's potential disagreement with them.

Table 2.1 Transformations in Heidegger's Word Usage after *Being and Time* *

Earlier usage (source)	Later usage (source)	Reference
aletheia as unconcealedness (*BT*)	*aletheia* as clearing (*CP*)	Schalow: 44, 86
anxiety (*BT*)	startled dismay (*CP*)	Schalow: 341
apprehending (*BT*)	being-historical thinking (*CP*)	Schalow: 352
authenticity (*BT*)	insistence/inabiding (*CP*)	Polt: 127, Schalow: 182
authenticity & care (*BT*)	sheltering (*CP*)	Polt: 200
being-in-the-world as *Dasein*'s fundamental ontological structure (*BT*)	abiding (mid-1940s) → dwelling in the fourfold (late 1940s)	Schalow: 136, Mitchell: 296
between, the (*CP*)	the dimension (late 1940s & 1950s)	Mitchell: 119
Dasein (*BT*)	*Da-sein* (*FCM*, *CP*) → mortals (late 1940s & 1950s)	Schalow: 71, 350; Mitchell: 211
disclosedness of being-in (*BT*)	lighted clearing of *Dasein* (1930s)	Schalow: 86
disposedness (*BT*)	grounding (or "fundamental") attunement (*FCM*, *Mindfulness*)	Schalow: 236–7, *Mindfulness*: 276
ecstases (*BT*)	transporting (*CP*)	Polt: 166
existence + facticity + falling (*BT*)	dwelling (*BDT*)	Schalow: 110
existential task of *Dasein* (*BT*)	task of overcoming metaphysics (*CP*, "Overcoming Metaphysics")	*CP*: 5
falling (*BT*)	being-away (*CP*)	Polt: 143–4
hermeneutic phenomenology of *Dasein* (*BT*)	meditation on the event of appropriation as the happening of be-ing (*CP*)	Polt: 1
language as element of disclosedness (*BT*)	language as the clearing-concealing abode of being itself (*CP*)	Schalow: 207–8
ontological difference (*BT*)	the interval of "the between" which allows that differentiation to occur (*CP*)	Schalow: 73, 106, 182
phenomenology (*BT*)	"the other beginning" of thinking (*CP*)	Schalow: 273
question of the meaning of being (*BT*)	question of the truth of be-ing (*CP*)	Schalow: 306, Polt: 46
standing-out/(the ecstatic) (*BT*)	standing-in [the clearing of being]/the ek-sistence of human beings ("Letter on 'Humanism'")	Mitchell: 214

Earlier usage (source)	Later usage (source)	Reference
temporality & spatiality (*BT*)	time-space (*CP*)	Schalow: 356, *CP*: 28
temporality of *Dasein* (*Temporalität*) as the "mirror image" of the temporality of being (*BT*)	the truth of being, or its sheltering unconcealment through the event of appropriation ("enowning") (*CP*, *The Event*)	Schalow: 346
time as the horizon for any understanding of being and for every way of interpreting its meaning (*BT*)	time as the clearing of the self-concealing of presence (*On Time and Being*)	Schalow: 354
transcendental knowledge as an investigation of how *Dasein* reaches out to a world and to being, making possible its experience of beings (*BT*)	the end of transcendental thinking because: (1) *Dasein*, as a "thrown thrower," can never completely control the happening, i.e., it cannot have priority over being and (2) be-ing does not ground beings, i.e., be-ing is not beingness (*CP*) AND the in-between (*Mindfulness*)	Polt: 46–8
understanding (*BT*)	language as the belonging together of being and human being through the directive of the event of appropriation ("enowning) (*CP*, "Letter on Humanism")	Schalow: 367
world (ontical, *Dasein*'s) (*BT*)	world as the open expanse, clearing, and play of time-space (*CP*)	Schalow: 393–4
world-poor (*FCM*)	worldlessness (*CP*)	*CP*: 218

* In the first two columns are found Heidegger's earlier and later word usages, respectively, followed in parentheses by exemplary texts of Heidegger's. In listing these transformations, I do not distinguish between outright replacement of one concept with another and a more modest modification in conceptual understanding. In a few cases, decades are listed to point toward multiple sources in Heidegger's work during that period. In the third column are found citations for the source of the claim about particular word usages, although in three cases Heidegger's work itself is cited. *BT*: *Being and Time*=Heidegger 1962; *FCM*: Heidegger 1995; *CP*: *Contributions*=Heidegger 2012b; "Overcoming Metaphysics": Heidegger 2003a; *BDT*: Heidegger 2013b; "Letter on 'Humanism'": Heidegger 1998b; *Mindfulness*: Heidegger 2016c; *The Event*: Heidegger 2013d; *On Time and Being*: Heidegger 1972a. Schalow: Schalow 2020; Polt: Polt 2006; Mitchell: Mitchell 2015.

Let us briefly consider some transformations listed in Table 2.1 that suggest (at least to some people) the possibility of greater affiliation between human *Dasein* and our fellow animals:

Authenticity and care (*BT*) → sheltering (*CP*). In *Being and Time*, care is described by Heidegger as a preliminary understanding of the being of human *Dasein* (*Zeitlichkeit*). It may be either authentic or inauthentic. As authentic, care implies disclosedness (articulated as: disposedness, understanding, and discourse) that manifests as self-disclosure, world-disclosure, and the discovery of entities. As inauthentic, there is still disclosure, but it is to *das Man*—the anyone-self—and not to an authentic self. The focus on human *Dasein* and the various articulations indicate Heidegger's human-centric understanding. In *Contributions*, disclosedness of *Dasein*'s being is placed in tension with a *sheltering* of being that serves to protect and preserve its mystery. Now, *Dasein*'s being can be disclosed while still sheltered (i.e., its mystery remains), but this opens up the reasonable possibility that the being of our fellow animals (and that of all entities) can at least be sheltered. However, the "sheltering" of animal being is not the same as *Dasein*'s concealedness of being, which is co-primordial with its unconcealedness (see Withy 2022). While the argument is complex, sheltering may be more closely related to what Withy calls being's "self-concealing." Regardless, it offers no hint that our fellow animals, too, are (can be) in *Dasein*.

Being-in-the-world (*BT*) → abiding (mid-1940s) → the fourfold (late 1940s) and, relatedly, *Dasein* (*BT*) → *Da-sein* (*FCM*, *CP*) → mortals (*CP*). In *Being and Time*, being-in-the-world is *Dasein*'s way of being. In *FCM*, we have seen that our fellow animals are world-poor, so being-in-the-world and *Dasein* appear as human-centric terms. Later on, "the fourfold" represents the primordial unity of earth, sky, divinities, and mortals. These terms are not meant to be understood naturalistically (and a deeper discussion must wait), but *Da-sein*—and not inauthentic *Dasein*—is more closely related to "mortals." However, our fellow animals are not mortals in Heidegger's sense, in part because they do not die; instead, they perish. They do not fit "in" any of the fourfold's moments except to the extent that they matter to *Da-sein*, where they (ontologically speaking) might then be considered—within a unitary understanding of be-ing—as earth's nourishment of mortals, but such an understanding is an unwarranted anthropocentrism.

Question of the meaning of being (*BT*) → question of the truth of being (*CP*) and, relatedly, Temporality of *Dasein* (*BT*) → the truth of being or its sheltering unconcealment through the event of appropriation (*CP*, *The Event*). According to Heidegger, the meaning of being is a question only for *Dasein*, and not for our fellow animals. However, the unconcealment of being through the event of appropriation might seem open to all who have experiences, including our fellow animals. After all, it seems counter-intuitive to argue that our fellow animals are not open to the presence of entities. But Heidegger would counter that they are not open to entities *as entities*. For Heidegger, having experiences entails having an understanding of life's meaningful structures, and such understanding is missing from our fellow animals.

Transcendental knowledge (*BT*) → the in-between and the end of transcendental knowledge (*CP*). Transcendental knowledge is acquired through the disclosedness of being—particularly one's own—and through worlding. Transcendence links being and beings—Heidegger's ontological difference. Later, Heidegger stops referring to transcendence because he considers the term as too indebted to psychological and metaphysical thinking. We might wonder whether avoiding "transcendence" could open up possibilities for our fellow animals, but that depends on what replaces it. In *Contributions*, Heidegger writes about the "in-between"—literally, the between of being and beings—in order to emphasize both the unresolved tension between the two and the importance of the in-between itself. While this becomes amplified as a significant shift in Heidegger's thinking, its implications for our fellow animals remain obscure.

world (ontical, *Dasein*'s) (*BT*) → world as the open expanse, clearing, and play of time-space (*CP*). In *Being and Time* (p. 93), Heidegger describes four understandings of "world" but thereafter in *Being and Time* prioritizes "world" as the ontical (factical) world of *Dasein*. Heidegger does not consider this world with respect to our fellow animals. In *Contributions*, Heidegger uses different language for understanding the world—the open expanse, clearing, play of time-space—and we might wonder whether this offers our fellow animals the possibility of having a world. After all, Heidegger shifts the emphasis from *Dasein*'s being to being as understood by *Dasein*. But then again, there remains no understanding (and disclosure) except as *Dasein*. In *Contributions*, the "open" is a formal indication of the site of being's appearance, where *Dasein* can encounter entities through the "play of time-space." The "clearing" of being is the *Da* ("there") of *Da-sein* ("being-there"), the grounding abyss (un-grounded ground) of truth (*aletheia*) itself. In this clearing, the being of entities is revealed. None of this encourages the thought that our fellow animals can have a world.

Now let us consider two transformations from Table 2.1 that clearly reinforce the isolation of human *Dasein* from our fellow animals:

apprehending (*BT*) → being-historical thinking (*CP*). Apprehending is the simple phenomenological awareness of something in its presence, as opposed to "thinking" (*denken*), a term that is avoided in *Being and Time*. Apprehension is central to Heidegger's interpretation of being in that it presupposes *Dasein*'s being-in-the-world. It would seem to be difficult to argue that our fellow animals are closed to apprehending, but Heidegger thinks that this is the case, ontologically speaking. For example, the title to §59b of *FCM* includes the phrase, "captivation as the animal's having any apprehending of something as something withheld from it" By the time of *Contributions*, being-historical thinking is central to understanding being, which is now called historical being, or be-ing. As the name suggests, this sort of thinking incorporates a rich sense of the past, but also of the future and an openness to destiny. Heidegger places his discussion of being in the context of its emergence in pre-Socratic thinking, followed shortly thereafter by its withdrawal from subsequent Western thought—beginning with Plato and his understanding of being as the beingness of entities. In his discussion of philosophers, Heidegger leaves little space to fit our fellow animals into being-historical thinking.

language as element of disclosedness (*BT*) → language as the clearing-concealing abode of being itself (*CP*). In *Being and Time*, disclosedness is a fundamental ontological term (see List 1.1) for *Dasein*'s openness to being and beings. Heidegger recognizes four structural moments of disclosedness of everyday *Dasein*, one of which is discourse (together with disposedness, understanding, and fallenness). Discourse is the ontological foundation of language, which allows *Dasein* to articulate being-in-the-world. By the time of *Contributions*, language as an existentiale has been elevated from one of the elements of disclosedness to that which allows unconcealment as such to take place. Language is no longer a direct expression of *Dasein* but rather is to be found in sheltering of what is not said, i.e., the clearing-concealing abode of being itself, or "the house of being." Heidegger's understanding of language as it relates to being indicates that *Dasein*'s being cannot be that of our fellow animals.

Now, let us turn to some of Heidegger's individual post-*FCM* works:

The Essence of Truth. On Plato's Cave Allegory and Theaetetus

(lectures of 1931–1932; Heidegger 2013c)

Heidegger says in these lectures on Plato's allegory of the cave that the cave dwellers comport to the cave's shadows as unhidden beings while an animal could never do this. Absent comportment, our fellow animals are unable to perceive beings as *un*hidden. Human *Dasein* exists—in Heidegger's special sense—by being open to the unhiddenness of beings in their totality ("as a whole"), which necessarily occurs through a confrontation between being and beings. By contrast, plants are "closed in upon" themselves, animals are "restricted … in their environment," and for the stone, it simply "occurs" (p. 62).

While perception of beings comes from "seeing," for seeing to occur there must be a relationship (a "yoke") between what is merely visible—a sensory perception that would appear to be accessible to humans and our fellow animals alike—and a "higher seeing," a synthesis of sensory perception that is open only to human *Dasein*. But, Heidegger (2013c: 187) says,

> The *aisthesis* [perception] of man depends *essentially* on sense-organs and bodily constitution. But just for this reason the *bodily constitution* of man is fundamentally different to pure nature: it is primordially inserted in the striving for being. It is not the case that man is first an animal and then something else in addition. Man can never be an animal, i.e., can never be nature, but is always either over the animal, or, precisely as human *under it* (whereupon we say that man becomes like an animal). Since nature does not have the inner elevation of *existence* which belongs to being-human as *being out beyond oneself*, it is incapable of *falling*.

In light of this, we may doubt whether even *Dasein*'s sensory moment can be shared with our fellow animals. Here, "moment" is not to be confused with "step," as in a sequence of temporal events.

Introduction to Metaphysics

(lectures of 1935; Heidegger 2014)

These lectures further develop Heidegger's question of being from that presented in *Being and Time*, but now with a more urgent, political motivation that aims to open us to the possibility of reversing the historical trajectory of being's understanding. Our fellow animals are little mentioned, but there are a few brief statements worth noting beyond those already found in *FCM*, namely, that the animal has neither world nor environment (pp. 49–50) and that only "saying beings"—not our fellow animals—can, and must, understand being (pp. 90–2). Heidegger says that such an understanding requires distinguishing between being, not-being, and seeming (-to-be-what-it-is-not). Consequently, the human being finds him/her-self in constant "de-cision," not in the sense of having to make judgments—or not—but in the sense of openness to being—or not. Heidegger says that all humans think, unlike our fellow animals, but that not all humans are thinkers (p. 130). Rather, those who think being—the thinkers—are the rare philosophers and poets.

"The Origin of the Work of Art"

(Essay of 1935–6; Heidegger 2002b)

This essay makes two claims that each strongly separate human *Dasein* from our fellow animals. First, Heidegger (2002b: 23) says,

> The stone is world-less. Similarly, plants and animals have no world; they belong, rather, to the hidden throng of an environment into which they have been put. The peasant woman, by contrast, possesses a world, since she stays in the openness of beings.

Not only do our fellow animals have no world, but in this regard they appear to be lumped together with plants, and even stones.

The second comment (Heidegger 2002b: 46–7) of interest concerns language:

> The conveying of overt and covert meanings is not what language, in the first instance, does. Rather, it brings beings as beings, for the first time, into the open. Where language is not present, as in the being of stones, plants, or animals, there is also no openness of beings, and consequently no openness either of that which is not a being or of emptiness.

Language necessarily brings beings into the openness of being. But lacking language, our fellow animals have no access either to being or beings. In this regard, they are again in the same category as plants and stones.

Schelling's Treatise on "the Essence of Human Freedom"

(Lectures of 1936; Heidegger 1985)

Worldviews express cultural- and time-specific values about human life. After an early flirtation with worldview as a philosophical project, Heidegger had by

1919 come to see philosophy as a science. Later, he disavowed that claim too because science cannot investigate being but instead must assume it. Nevertheless, Heidegger remained interested in worldview, if only because it indicated, for him, a givenness about being that was unjustified. In his discussion of Leibniz's scientific worldview found in the treatise on Schelling, Heidegger (1985: 18) makes a critical distinction:

> Animal and plant, too, have their world view, better expressed, are a world view, a way in which world opens itself, although only mutedly and darkly. A world view belongs to the constitution of every being to the extent that it is thought monadically.

Here, the distinction is between "having" a worldview and "are" a worldview. Only humans can have a worldview, and only humans—not the other animals—can "see" that and what an animal worldview is.

Later in these lectures, Heidegger (1985: 140) says,

> The animal is an individual this. It could not be this if the craving for individuation were not in it. But the animal never comes to itself, in spite of this craving. It rather merely serves, again, the species in spite of this craving. And it could not perform this service if a universal will did not strive within it with the particular will at the same time.

Here, "will" refers to animal instincts. Whether an individual, a species, or an animal in general, no ontological distinction is to be made as regards the grounding of their instinctual behavior. Hence, Heidegger can say (as he did in *FCM*) that the being of all animals is captivation.

Contributions to Philosophy (of the Event)

(Written 1936–8; Heidegger 2012b)

Contributions is not just another one of Heidegger's books. After *Being and Time*, *Contributions* is considered his second groundbreaking effort at exploring being/be-ing,[1] and the in-between. While the overall aim of the book has little to do with our fellow animals as such, they are mentioned in multiple contexts, and these illuminate Heidegger's thinking. For one, Heidegger says that sheltering abides in the works of human *Da-sein*—for example, in certain works of art ("The Origin of the Work of Art": Heidegger 2002b)—but also in the manner in which we encounter non-*Dasein* entities, including our fellow animals (*Contributions*: 57). While sheltering is a way of being human *Da-sein*, it is not that for the animal, whose way of being captivated is "received" by *Da-sein*.

Later in *Contributions* (p. 218), Heidegger cryptically says the following about "life," although it could (and does) refer to animals as well:

> The *darkening* and the essence of *instinct*. The *preservation* of the self and the *priority of the "species"*; this does not know any "individual" as egotistical.

> The darkening and *worldlessness*. (Earlier as *world-poor!* Liable to be misunderstood. The stone not even worldless, because even without darkening.)

Here, "darkening" indicates the inability of our fellow animals to have access to a world, but also the ability of *Dasein* to imagine this darkening for the animal but not for the stone. Heidegger also says that all life (animals), even the most "lowly" (*Contributions*: 81), has a drive to self-preservation, albeit "instinctual," and that this drive is an absorption by stimuli, reflecting their worldlessness.

We have already discussed how biology plays a role in Heidegger's thinking about animal being, although Heidegger is never satisfyingly clear about how he makes the transition from observing animal behavior, for example, to claiming animal being in its essence. Even so, he is clear that biology by itself remains misguided. Heidegger (*Contributions*: 217) says,

> But must there be "biology," since it derives its justification and its necessity from the sovereignty of science within modern machination? Will not every biology necessarily destroy "living beings" and thwart a fundamental relation to them? Must this relation not be sought completely outside of "science"? In what space should this relation abide?
>
> "Living beings," like everything that can be objectified, will offer scientific progress endless possibilities and yet will also withdraw more and more, the more groundless becomes science itself at the same time.

Heidegger is arguing that biology is a servant of machination—a new form of being—and thereby has participated in the destructive transformation of authentic human beings, ontologically speaking. Later, Heidegger (*Contributions*: 317) says,

> Will the gods then be *over and done* and a relapse into the mere life of *world-poor* creatures commence, ones for whom the earth has always remained only something to be exploited?

Here the "world-poor creatures" refer to the technologizing human *Dasein* who are on the path to their own worldlessness, and not to our fellow animals.

I conclude this brief discussion with an important clarification that Heidegger makes in *Contributions* (p. 237).

> The talk of "human Dasein" (in *Being and Time*) is misleading insofar as it suggests that there might also be animal and plant "Dasein." … Rather, "human Dasein" signifies the uniqueness of that being, the human being, to whom alone Da-*sein* is *proper*. … Da-sein—the mode of *being* that is distinctive of humans *in their possibility; thus it is no longer at all necessary to add "human."*

Here, Heidegger is explicit that *Dasein* is not open to our fellow animals. To emphasize this, he introduces "*Da-sein*" as a more essential way of thinking "*Dasein*" in its possibility. Given that animals live in their actuality, and not their possibility, Heidegger says that "human *Dasein*" and "*Dasein*" are redundant terms. Heidegger's "elevation" of *Dasein* into *Da-sein*—and being into be-ing—incorporates a new understanding of historical contingency and fate in a way that leaves behind *Being and Time*'s more systematic and universalist approach to understanding the human being, but still, there is no basis here for claiming a change in his understanding of animal being.

Basic Questions of Philosophy. Selected "Problems" of "Logic"

(Lectures of 1937–8; Heidegger 1994)

I have already hinted at Heidegger's ambiguous attitude toward evolution and his decidedly negative attitude toward natural selection. I use my commentary on Heidegger's *Basic Questions of Philosophy* to introduce his and my views. Then in Chapter 3, I present a more in-depth discussion of evolutionary thinking. Heidegger (1994: 46–7) says,

> It is a well-known fact that the natural sciences admit a historiographical consideration of their own past merely as an addendum, since for them what is past is simply what is no longer. Natural science itself only deals with present nature.

Four claims can be identified herein. First, natural sciences operate within the framework that "… what is past is simply what is no longer." Here, Heidegger ignores the historical biosciences, most prominently evolutionary biology, where characters (morphological, developmental, molecular) present in distinct lineages of modern-day organisms are compared in order to make inferences about transformations in those characters over time and across lineages, and to infer what lives of (extinct) ancestors may have been like. For the evolutionary biologist, the past is continuous with and informative of the present, and it delimits future possibilities. Charles Darwin and Alfred Wallace illustrated this long before Heidegger was even born. So second, the idea that for natural historians the past is "merely as an addendum" is also false, as is (thirdly) the claim that the evolutionary biologist "… only deals with the present nature," and (fourthly) that all this "… is a well known fact."

How could Heidegger have gotten this so wrong? Perhaps he thought that biologists only worked within a highly reductionist, mechanistic framework, but then he would have had to ignore his prominent fellow countryman Ernst Haeckel (1834–1919), who published extensively from the 1860s into the twentieth century and who coined the term "ecology." Likely, he also thought that the numerous natural historians of his time (and of the preceding centuries) were not really scientists. Perhaps he agreed with the physicist Ernest Rutherford (1871–1937), who is reported to have said, "All science is either physics or stamp collecting" (https://www.newn.cam.ac.uk/wp-content/uploads/2018/06/Nadia-Mason-2.pdf). The extremeness of this statement is now acknowledged, but given that even more recent scientists sometimes misunderstand the role of evolution in biology (see account of J.D. Watson in E.O. Wilson 2019), Heidegger's own misunderstanding should not be surprising. He was probably simply following his favorite, contemporary biologist, Jakob von Uexküll (1864–1944), who was also explicitly anti-Darwinian (i.e., opposed to natural selection) and an ambiguous supporter of evolution itself (von Uexküll 2018: originally published in 1920).

Ponderings VII–XI: Black Notebooks 1938–1939

(Heidegger 2017)

In "Ponderings X" (Heidegger 2017: 220), Heidegger states (in one brief paragraph of 128 words) that his earlier characterization of stone, animal, and human being (in *FCM*) as a "kind of world-relation" (i.e., worldless, world poor, world forming, respectively) is "insufficient." Rather, it would be more proper to say "*fieldless and worldless* [the stone], / *field-dazed—worldless* [the animal], / and *world forming—earth disclosing* [the human]." The key new term—"field-dazed" (also translated as "captivated in its field")—is attached only to the animal. In "Ponderings VIII," Heidegger (2017: 100) hints at what I think he means by "field-dazed" when he says that "life" (including nonhuman animals but not humans) incorporates "a surrounding field (of beings) as a conditionality of the organism" even though its essence "would retain a rigidity. The living thing would lack an openness to beings as such." In this sense, field-dazed appears very similar in meaning to what he says in *FCM* about animal captivation and its encircling ring of disinhibitors.

How does Heidegger's new reconstruction work in his larger understanding? First, the animal is unambiguously to be considered worldless, and no longer world poor (with its double meaning) as presented in *FCM*. Doing this sediments his prior claim of the ontological abyss separating the animal from the human. Second, by introducing the "field-dazed" term, Heidegger maintains his previous distinction between the animal and the stone, even though both are now to be considered worldless. But is "field-dazed" any more understandable than "world poor"? Consider a rock glacier as a field of sometimes-moving stones. The rock glacier's behavior would seem to match quite well the above understanding of field-dazed. Obviously, Heidegger's main interest is human *Dasein*, not rocks (or even, I have argued, animals), so I think his aim in this new reconstruction is principally to make more definitive the claim of an abyss separating the human from the animal. But in a speculative mode, one could also say that Heidegger's distinction between field-dazed (animal) and fieldless (stone) is meant to push against the scientistic claim that there is no fundamental difference between the living (animal) and the nonliving (rock). In light of the brevity of comments concerning animals in these *Ponderings*, and absent "field-dazed" talk from Heidegger's other works, I see no evidence from these *Black Notebooks* of a fundamental change in his thinking from that presented in *FCM*. (See Pineda Saldaña 2025 for a partially dissenting view.)

Mindfulness

(Written 1938–9; Heidegger 2016c)

In *Mindfulness*, Heidegger tells us that *Da-sein* is not to be understood anthropologically or metaphysically, and "any attempt at grasping Da-sein predominantly or even exclusively with a view toward man remains inadequate" (p. 282). Instead, *Da-sein* is intermeshed with the human, the gods, and the mystery of being which conceals ("the earth") but also opens up ("the world") in what Heidegger calls (in one translation) the ap-propriating event. As ontological, this "event" must be thought in the unity of all its aspects. Given Heidegger's holism, we might question whether the previous characterization in *Being and Time* of authentic human beings as a *subset* of all human beings can be accurate. Instead, it seems that here Heidegger has shifted to considering human beings as *historical*, and only in their essence, that is, in their relation to be-ing. But with this move, the connection to the embodied human and our fellow animals becomes less clear. Heidegger (2016c: 154) says,

> Only that which is "historical" can also be "un-historical". The "un-historical" remains fundamentally different from that which is without "history" since that which is without "history" lacks history, for example, the animal and all forms of "life."

The "history" that Heidegger is writing about is not that represented in natural history, but rather a history that projects a future of one's very own that has emerged from one's past and which thereby is meaningfully/authentically lived in the present.

Elsewhere in *Mindfulness* (p. 111), Heidegger says,

> And where being *seems* to open itself up to a being, as in the animal, there everything is overlaid by the mere environing, which is called such, because it can never "give" anything to a being that is incapable of "taking" and "renouncing".

Our fellow animals may "have" an environment, but in Heidegger's terms, an environment, as opposed to a world, is a darkness that only humans can be aware of. Any claim of openness by our fellow animals to others or to their environment is simply anthropocentric thinking on the part of our fellow human beings, or so Heidegger would have us believe.

Interpretation of Nietzsche's Second Untimely Meditation

(Lectures of 1938–9; Heidegger 2016b)

The second of four essays in Nietzsche's *Untimely Meditations* is entitled "On the Use and Abuse of History for Life" (originally published in 1874; see Nietzsche

1997) and is a target of Heidegger's broadly conceived interpretation of historical being. For Heidegger, Nietzsche's writings, considered in their totality, represent the culmination of Western metaphysics, but what is highlighted in this book is Heidegger's own transition from thinking being as still tinged with metaphysics (Heidegger's understanding) to thinking "being historical" as post-metaphysical. Heidegger's use of "historical being" (and "being historical") is based on a different understanding of history than what Nietzsche (and others in the metaphysical tradition) uses in his second essay. Heidegger emphasizes the happening or the event of appropriation (or "adaptation") of a particular human *Dasein* to the presencing of beings, and vice versa. In thinking historical being, he refers to "*Geschichte*"—translated as "the past as a foundation of our reality"—and "*Geschictlich*"—translated as "historial in its essence." By contrast, Nietzsche emphasizes becoming as the beingness of beings and constancy as a recounting of the "facts" of human life. The German terms Heidegger uses when referring to Nietzsche are "*Historie*"—translated as "historiology"—and "*historisch*"—translated as "historical." Heidegger says that Nietzsche's understanding of the historical, and what it implies for an understanding of life, is not grounded in the truth of being but rather is misconceived as a "biologism." By this, he means two things: First, Nietzsche considers biology as providing the most basic determination of beings as a whole, and by "biology" Heidegger here means a biology of human culture. And second, Nietzsche's thinking falls broadly within the Aristotelian legacy, now updated by more modern scientific thinking, to consider the human as *animal rationale*; that is, as primordially an animal with the capability for rationality added on, contra the abyss that Heidegger considers to separate the essences of the human and our fellow animals, an abyss that science fails to see given its exclusive focus on beings rather than being. This failure does not simply result in an absence of understanding but in a *mis*understanding of the human and our fellow animals alike.

Nietzsche argues that our fellow animals, unlike humans, do not respond to a past or a future, and thus must be considered "unhistorical." In response, Heidegger makes three central points. First, he rejects the idea that our fellow animals are unhistorical because this implies that humans and our fellow animals share a deeper unity which allows the comparison (i.e., historical/unhistorical) in the first place. This deeper unity would be the unhistorical, out of which humans—and their historicality—emerge from animality. Second, he says that our fellow animals are "without history," which is not the same as "unhistorical," because being without history allows only a contrast to be made with the human, not a comparison. And third, Heidegger challenges even the thought that our fellow animals can "make present" as such, that is, one in which they can be aware of beings as beings. Of course, through disinhibition they can still respond to other beings in their milieu (i.e., their ring of disinhibition), but this remains a lack of relation to those beings as such—a captivation. "Milieu" is Heidegger's carefully chosen word to reference that which "surrounds" the animal. "Milieu" is neither the "environment" that biologists misguidedly talk about (according to Heidegger) nor the "world" that Heidegger attributes exclusively to the human being.

Heidegger adds that the stone neither comports—like the human—nor behaves—like our fellow animals. The stone cannot even be captivated. Heidegger is saying that while our fellow animals relate to other beings, although not to those beings *as* beings, the stone does not relate to anything at all. But the commonsense inertness of the stone that Heidegger uses to exemplify the nonliving seems conveniently chosen to be distant from more obviously dynamic, chemistry-rich, nonliving things (e.g., consider the sun or moon, a volcano, our atmosphere) which interact with other entities, including ourselves, in specific and complex ways.

According to Nietzsche, the distinction between the unhistorical and the historical is likewise reflected in the difference between "constant forgetting" (or simply "forgetting") and "always having to remember again" (or simply "remembering"), but Heidegger says that this too provides no basis for distinguishing our fellow animals and the human, since even forgetting implies the possibility of retaining what belongs to it, a possibility that Heidegger denies to our fellow animals. Both forgetting and remembering are part of the human essence, and both are in the truth (and untruth) of be-ing.

The History of Beyng

(Written 1938–40; Heidegger 2015)

In *The History of Beyng* (among other works), temporality as the being of human *Dasein* is transformed through his understanding of the event of appropriation (adaptation) and of history (*Geschichte*). This history applies only to those few humans who have become *Da-sein* by preparing themselves for the possibility of the event. Humans who have not prepared themselves remain unhistorical. Clarifying his thought that animals are without history, he says—in modern terms—that biological inheritance and evolution cannot ground history as *Geschichte*, in part, because animals are seen as being exclusively about past happenings (Heidegger 2015: 81). By contrast, Heidegger's history is about the future, about the de-cision which has yet to be made concerning be-ing itself. But, one may ask, do biologists claim that inheritance and evolution are only, or even primarily, about the past? Can scientific claims be rephrased in terms that are more Heidegger friendly? Is Heidegger preventing by fiat a history that might be informative of be-ing? But second, Heidegger is also saying that history is the event of an individual human *Da-sein*, not *Dasein* in general. Can there even be such a history for our fellow animals? Well, as long as we treat our fellow animals as a group, then this is likely not possible. But is it necessary that we do this? Can we think of an individual animal's inheritance, or behavior, as attuning him/her to a world of possibilities? And third, Heidegger rejects there being a history of occurrent (non-*Dasein*) entities, simply claiming that our fellow animals, in particular, can have no involvements. This would be a serious claim if it were true. But is it? We consider these matters further in Chapters 3–5.

The Event

(Written 1941–1942; Heidegger 2013d)

Heidegger (2013d: 10) says,

> A *being* is a *possible* object, something standing over and against, only because it stands in the open domain of being. … And precisely this open domain is denied to plant, animal, and everything that merely lives.

Parmenides

(Lectures of 1942–3; Heidegger 1992b)

A seemingly unlikely venue for Heidegger to discuss animals and biology, his *Parmenides* lectures actually offer important insight into his view a dozen years post-*FCM*. The lectures themselves aim to make his listeners aware of the numerous ways truth as *aletheia* (and *létha*)—translated as "unconcealedness" (and "concealedness" or "oblivion")—has been misunderstood since the time of Parmenides, and what is at stake in these differing presumptions. Like *FCM* (and *Contributions*), *Parmenides* is a book of prophecy. And in both books, our fellow animals play a contrastive role to humans.

For example, when discussing Pindar (in §5c), Heidegger says that the word is connected to the essence of the human hand, not as a body part (or not simply as such) but as the unity of presencing and dwelling through the reach of the hand. To emphasize the remoteness of biological concerns from his aim, Heidegger (1992b: 80) says,

> The hand exists as hand only where there is disclosure and concealment. No animal has a hand, and a hand never originates from a paw or a claw or talon. … The hand sprang forth only out of the word and together with the word. Man does not "have" hands, but the hand holds the essence of man.

As in *FCM*, Heidegger is here talking holistically and essentially, that is, with reference to being. Reductionist approaches (e.g., biomechanics and physiology) which emphasize the overall similarity of animal hands, including that of the human, are not based on essential thinking; that is, they cover up any understanding of the distinct essences of the human and our fellow animals. "The hand entrusts to the word the relation of Being to man and, thereby, the relation of man to beings" (Heidegger 1992b: 84).

Another reference to our fellow animals comes in a discussion of looking (§6e), which Heidegger says is denied to animals. Thought essentially, looking does not refer to a physiological process; rather, it is a consequence of *aletheia*. For the human, looking is about encountering being. Heidegger says that this is extraordinary in that it is beyond an encounter with beings. He says this is "uncanny," a fundamental mood in which the human self experiences his/her own *un*familiarity of being-in-the-world, the sort of removal from the world of ordinary things that Heidegger references with respect to anxiety in *Being and Time* and contemporary boredom in *FCM* (see Withy 2015). Uncanniness cannot be understood biologically, but only as emerging from looking. Still, Heidegger (1992b: 107) does say, "the animal's 'look' can concentrate in itself a special power of encounter." Heidegger explains that an animal's "encountering look" is distinct from the human's "grasping look" in that only the latter is a disclosure. Once again, our fellow animals only *appear* to do what humans do, namely, look.

Heidegger (1992b: 107) also says, "the animal has a peculiar intermediate position," presumably between the human (with the grasping look) and the stone (with not even the encountering look). Heidegger's presentation of this intermediate position as ontological conveniently fits with an ontical account of animal instinct. Unlike the stone, many animals have eyes and a brain, making it hard to deny their receptivity (i.e., through the encountering look) and responsiveness to that which is outside themselves. But unlike the human, the animal does not (according to Heidegger) see things *as* the things they are (i.e., through the grasping look).

At the end of *Parmenides*, Heidegger discusses "The Eighth Elegy" in the *Duino Elegies* by Rainer Rilke (Rilke 2000). Rilke, too, writes about the open, but now it is our fellow animals who experience this, and not humans. Rilke's elegy begins:

> With all its eyes the creature-world beholds the open. But our eyes, as though reversed, encircle it on every side, like traps set round its unobstructed path to freedom. What *is* outside, we know from the brute's face alone; for while a child's quite small we take it and turn it round and force it to look backwards at conformation, not that openness so deep with the brute's face. Free from death.

Heidegger calls Rilke's understanding a biologism, in which the open—as understood by Rilke—remains focused on beings to the very oblivion of being. Furthermore, Heidegger says that Rilke contrasts the animal ("the creature") to the human on the basis of metaphysical beliefs about a fundamental distinction between the a-rational living being and the rational living being. Yet, Heidegger says, how can the very being who has words be denied unconcealedness, while the one without words has access? Heidegger (1992b: 159) concludes that Rilke is not referring to "the open in the sense of the unconcealedness of the unconcealed" but rather to a limitless spatial realm that is conventionally understood as nature with all its causal interactions. Interpreting Rilke, Heidegger (1992b: 157) says that humans are forced to confront objects, while "the animal's gaze is not trammeled by any objects but can go on infinitely, in some unknown way, into the objectless." Furthermore, animals are "free from death." Interpreted within the frame of Rilke's biological metaphysics, Heidegger concludes that the animal's "freedom" versus the encapsulation of the human results from a prioritization of the unconscious over the conscious, and that the essence of the human is conceived on the basis of the essence of the animal. In countering Rilke, Heidegger is steadfast that only humans see the open, and he thinks that Rilke's poem diminishes all living beings because "the mystery of the living being," namely, being itself, goes unheeded.

Heidegger (1992b: 160) also repeats his claim made in *FCM* that the animal (and the plant) is essentially different from the stone. First, he says that the animal "is related to"—importantly, he does not say, "relates to"—"his circle of food, prey, and sex," and has a "peculiar arousal of excitability" which is "suspended in something outside [itself] without ever being able to 'see' the outside or the inside, i.e., to have it stand as an aspect unconcealed in the free of Being." A key phrase here is "peculiar arousal of excitability," which may be related to the "proper peculiarity" that he assigns to the essence of animals in *FCM* (see my discussion of *FCM*'s §56 in Chapter 1, Section D). Regardless, the meaning of both phrases remains obscure, and while the ontical references to food, prey, and sex are *differences* from the stone—at least as our use of language has incorporated these differences—it is not clear that they support essential ontological differences. All Heidegger says about the stone is: "And never would it be possible for a stone ... to elevate itself toward the sun in jubilation and to move like a lark, which nevertheless does not see the open" (Heidegger 1992b: 160).

"Letter on Humanism"

(Essay of 1946; Heidegger 1998b)

Five years after *The Event* and sixteen years after *FCM*, Heidegger (1998b: 246–7) continues to warn against viewing human essence in the same way as animality.

> Above and beyond everything else, however, it finally remains to ask whether the essence of the human being primordially and most decisively lies in the dimension of *animalitas* at all. Are we really on the right track toward the essence of the human being as long as we set him off as one living creature among others in contrast to plants, beasts, and God? We can proceed in that way; we can in such fashion locate the human being among beings as one being among others. We will thereby always be able to state something correct about the human being. But we must be clear on this point, that when we do this we abandon the human being to the essential realm of *animalitas* even if we do not equate him with beasts but attribute a specific difference to him. In principal we are still thinking of *homo animalis* ... Such positing is the manner of metaphysics. But then the essence of the human being is too little heeded and not thought in its origin, the essential provenance that is always the essential future for historical mankind. Metaphysics thinks of the human being on the basis of *animalitas* and does not think in the direction of his *humanitas*.

By linking human essence and animality, so Heidegger says, neither is thought in their (ontological) origins. Such origins are not of the temporal sort that evolutionary biologists invoke when studying beings but rather indicate a primordiality of being itself. The only particular being for whom this matters is *historical* (human) being, not human beings metaphysically conceived. Historical being is ecstatic; that is, it is futural and "beyond itself," something that is opaque to human beings considered metaphysically and that is completely closed to our fellow animals. But if *humanitas*, not *animalitas*, is our origin, then how did we get there? According to Heidegger, the human is in *humanitas* insofar as he/she is ek-sistent, as opposed to mere existence (i.e., present-at-hand) or insistence (i.e., immersed in beings). But beyond this, *humanitas* is separated by an abyss from *animalitas*; that is, there is no way to bridge the two. Since Heidegger is not writing here about beings, but being, science offers no direct rejoinder. So Heidegger's rejection of the human being as *animal rationale* is not based only on the denial that reason is of the essence of the human being, as though reason is something that is simply tacked onto a preexisting animal body; rather, we must *begin* with *humanitas* if we are to understand human reason itself, and then only in a derivative sense. But must we accept Heidegger's final word as our own? Does it make sense to us as it must have to Heidegger?

To avoid confusion in a related matter, Heidegger (1998b: 247) does not deny embodiment, but he restricts the body to the human and says that its understanding cannot be based on science.

> The human body is something essentially other than an animal organism. … The fact that physiology and physiological chemistry can scientifically investigate the human being as an organism is no proof that in this "organic" thing, that is, in the body scientifically explained, the essence of the human being consists.

Again, the key point to understand is that Heidegger is thinking of the human body in its essence, that is, its being. In *FCM*, Heidegger already made a parallel distinction with reference to "organism." He said that if we think of the organism as the morphological unity of the animal, then we have failed to understand its essence (*FCM*: 258). Rather, the organism should be thought as "the capability for behavior in the unity of captivation" (*FCM*: 257). I will have more to say about Heidegger's understanding of embodiment when we consider the *Zollikon Seminars*.

"Introduction to 'What Is Metaphysics?'"

(Essay of 1949; Heidegger 1998a)

Heidegger uses "existence" (and "ek-sistence") in a special way that refers only to human beings (see "*Dasein and Da-sein*" entry in my Chapter 1, Section A). This allows Heidegger (1998a: 284) to say:

> The human being alone exists. Rocks are, but they do not exist. Trees are, but they do not exist. Horses are, but they do not exist. Angels are, but they do not exist. God is, but he does not exist. … The proposition "the human being exists" means: the human being is that being whose Being is distinguished by an open

> standing that stands in the unconcealedness of Being, proceeding from Being, in Being. The existential essence of the human being is the reason why human beings can represent beings as such.

Even though the later Heidegger shifts his earlier thinking of being as *Dasein*'s being toward less *Dasein*-centric ways of expressing being itself, it is clear that *thinking* being remains with human *Dasein*, and only human *Dasein*. But for me, the question remains: Can our fellow animals think being too?

"The Thing"

(Lecture of 1949; Heidegger 2013f)

and

"Building Dwelling Thinking"

(Lecture of 1951; Heidegger 2013b)

In "The Thing," Heidegger replaces the framework of being-in-the-world with a new ontological framework that speaks of dwelling in the oneness of the fourfold of earth and sky, divinities and mortals. Whereas in *Being and Time*, where an ontological understanding of being-in-the-world immediately leads us to focus on that unique being—human *Dasein*—who can have a world, in "The Thing" the ontological understanding of the fourfold is more about: (1) the world—and being—itself, (2) the essence of "things" in that world, (3) how we humans come to know things, but (4) how we are commonly deceived and, thus, do not know things as things. While humans are decentered within the fourfold, they remain uniquely necessary for thinging to take place. We might wonder where our fellow animals fit into the fourfold and thinging, and whether Heidegger has moved beyond the anthropocentrism of *FCM*.

When dwelling in the fourfold, mortals care for the things disclosed in their world by saving the earth, receiving the sky, and awaiting the divinities. The terms of the fourfold, like other ontological terms, are not to be thought as referencing a scientific reality or parts of a whole. But still, Heidegger (2013f: 176) is explicit as to the identity of the mortals:

> The mortals are human beings. They are called mortals because they can die. To die means to be capable of death as death. Only man dies. The animal perishes. It has death neither ahead of itself nor behind it. … As the shrine of Nothing, death harbors within itself the presencing of Being. As the shrine of Nothing, death is the shelter of Being.
>
> Metaphysics, by contrast, thinks of man as *animal*, as a living being. Even when *ratio* pervades *animalitas*, man's being remains defined by life and life-experience. Rational living beings must first *become* mortals.

So, a first point is that in Heidegger's new ontological schema, a human being can be viewed as having its own moment—called "the mortals"—within a fourfold of moments that are nevertheless a unity. When we view a human being in this way, we are thinking ontologically. (It is important to emphasize here that while all mortals are human beings,—viewed ontologically—Heidegger does not say that all human beings are mortals.) The unity of the fourfold is not a static unity, like some universal philosophical schema; rather, Heidegger (2013f: 178) calls it "the fouring." To be a mortal is "to be" this dynamism that perfuses world and the things therein. Simply said, the moments of the fourfold are ontological constructs of Heidegger. They—like all of Heidegger's ontological terms—should be thought of as formal indications, and not as definitions or instantiations.

A second decisive point derived from the above quotes is that our fellow animals are not mortals. So, where do they fit into the fourfold? I think the answer is that when they are considered as the organisms of biologists, they simply do not belong. Their being is differently understood: as mechanism but more recently as standing-reserve, for example, as model organisms for making generalizations or as "test objects" for quality-control purposes. But where do our fellow animals fit when their being is thought in terms of their "proper peculiarity with the capability to create organs" as described in *FCM* (see §§51,57)? Interestingly, Heidegger (2013f: 176) does not say that they are in the "earth" "category"; instead, he says, "Earth is the building bearer, nourishing with its fruits, tending water and rock, plant and animal." But, if the earth tends the animal, then where is the animal? Not with mortals or the earth, but also not in the sky and certainly not with the divinities. We have a conundrum, but that is because we are not thinking like Heidegger. To repeat, for Heidegger, animals are not material entities simpliciter. Like humans, they are ontical nodes in the world's "relational" web, but unlike humans, they are neither world forming nor selves, and their web is confined to a "disinhibiting ring" of which they are unaware. Animals become things—transiently concretized nodes of relationality—only when they are "thought" by humans within the context of the fourfold. Importantly, Heidegger does not refer to human beings as things; rather, they are always "who."

Two years after "The Thing" essay, Heidegger continues to develop his idea of the fourfold in the "Building Dwelling Thinking" essay. As a summary statement, we can say that it is only when speaking ontologically that earth and sky nourish our fellow animals. But further, since our fellow animals, like all other living, non-*Dasein* entities (e.g., plants, fungi, bacteria), are not assigned to a moment in the fourfold—unlike human *Dasein* when viewed as "the mortals"—I would say that our fellow animals have no *necessary* relationship to the fourfold. As an example, if we consider some animal that lives remote from the human gaze, then that animal simply does not relate to the fourfold. Our fellow animals can dwell in the fourfold only to the extent that they are in human *Dasein*'s world, for example, as a pet, a "stock" animal, or minimally a focus of human observation. Given the unity of the fourfold, "the mortals" do have a necessary relationship to the other moments, but only in their freedom. For example, the divinities ap-propriate human *Dasein*. Heidegger's own wording—"earth and sky, divinities and mortals"—suggests that the mortals have a particularly close relationship with the divinities. Nevertheless, he is averse to drawing sharp distinctions. Heidegger (2013b: 177–8, 180) says,

> The united four are already strangled in their essential nature when we think of them only as separate realities, which are to be grounded in and explained by one another.

But still,

> Men alone, as mortals, by dwelling attain to the world as world. Only what conjoins itself out of world becomes a thing.

For Heidegger, "conjoining" is abiding in what-being, and giving rise to things. As this is his concluding comment for the entire essay, it is clearly important to Heidegger that we, the readers, not forget that only for humans can a world *as* a world become manifest (or perhaps now better said: only humans can world and can thing), just like he claimed twenty years earlier in *FCM*. But the world itself, as clarified through Heidegger's understanding of "fouring," connotes a different understanding than the formal definition provided in *FCM*. First, there is a shift away from the *Dasein*-centrism of *FCM*, but not in the sense that now humans and our fellow animals are thought to be similar. Rather, Heidegger says that he has always been thinking ontologically, and that dwelling itself cannot be "reduced" to *Dasein*'s dwelling alone, let alone to some sort of biologism (as Heidegger would say). Nevertheless, it remains the case that without human *Dasein*, there can be no dwelling.

Second, there is a shift away from ahistorical thinking implicit in "world's" earlier formalism and toward "being historical" thinking.

As "The Thing" makes clear, it is not only mortals who dwell in the fourfold. Dwelling in the fourfold is a kind of residing—Heidegger (2013b: 175) refers to the "presencing of nearness"—but this residing is not a residing by some autonomous agent. Rather, earth and sky, mortals and divinities all dwell, not separately but together, and in so happening a world is manifest. We might wonder whether humans, conceived metaphysically, can have a world, given that they are "outside" the fourfold. And I think Heidegger's answer must be yes, they do have a world, but it is a different world, for example, the world viewed objectively and scientifically.

The fourfold is Heidegger's new attempt to think what it means to have a world. Near the end of "The Thing" essay, Heidegger (2013b: 180) says, "Things, each thinging from time to time in its own way, are heron and roe, deer, horse and bull." But these things—namely, our fellow animals—are things only where there is already a world understood as a world, that is, as human *Dasein*'s world. *FCM* left us with some ambiguity as to whether animal being (i.e., captivation) was merely a human perspective. The same question can still be asked about the fourfold, despite the fact that there has been a decentering of human *Dasein*.

"Language in the Poem. A Discussion on Georg Trakl's Poetic Work"

(Essay of 1952; Heidegger 1971a)

Georg Trakl was one of a few poets who attracted Heidegger's serious attention. Heidegger says of him: "his poetic work is historical in the highest sense. His poetry sings of the destiny which casts mankind in its still withheld nature—that is to say, saves mankind" (Heidegger 1971a: 196). Heidegger's essay—"Language in the Poem"—is "a dialogue between *thinking* and poetry" (Heidegger 1971a: 160), where the "thinking" is Heidegger's discourse on the poetry of Georg Trakl. While Trakl's poetry and Heidegger's essay on it focus on humankind, our fellow animals are amply cited. But understanding their role is neither straightforward nor unambiguous.

In my own reading of Heidegger's essay, we might first wonder what Trakl himself is saying; that is, we could confine our analysis (at least in part) to Trakl's poetry and other aspects of his life that inform us about his poems. But this is not my primary objective. Second, there is Heidegger interpreting Trakl's poetry—that is, Heidegger trying to understand what Trakl himself is saying. Likewise, this is not my primary aim. And third, there is Heidegger himself, with his own agenda, trying to place Trakl's words in the context of that agenda. For purposes of this book, this is what principally concerns me—Heidegger's agenda via-à-vis Trakl's poetry. But any attempt to talk about Heidegger's own agenda confronts the difficult challenge of disambiguating these three perspectives of interpretation. In despair, one might simply ignore this essay, but that would not be wise, for our fellow animals—and plants and stones[2]—are frequently cited in the quotations of Trakl's poetry that Heidegger selects, and we might wonder why and how Heidegger responds to them. Furthermore, Andrew Mitchell (2011, 2015) has interpreted this essay to indicate that Heidegger underwent a profound shift in his thinking about the being of our fellow animals, away from the captivation expressed in *FCM* and toward "exposure to world." In this segment, I provide my own analysis, which is not in complete agreement with that of Mitchell. But first, some context for this essay.

Heidegger was 62 years old when he wrote this essay. He had long since made the case for moving away from metaphysical being and toward historical being, but expressing this in standard philosophical vocabulary had proved inadequate for him. Instead, he came to reimagine philosophy as "thinking"—as thought provoking, not expository—and he seriously engaged painting, sculpture, and poetry in his writing, which offered new—and less formal—avenues of expression. In "The Origin of the Work of Art," Heidegger (2002b) says, "Poetry is the saying of the unconcealment of being." Of the various forms of discourse, poetry and thinking are uniquely able, so Heidegger says, to disclose both what is sayable, namely, the unconcealment of beings, and what is unsayable, namely, historical being itself. By leaving unsaid what is unsayable, poetry moves outside the world of entities and becomes open, even if only in "resonance" (see *Contributions*: 137), to the nothing that "is" being. However, having just said this, we could add yet another difficulty to understanding "Language in the Poem," which is that Heidegger himself thought that a poetic statement (such as Trakl's poetry) could only enter into a true dialogue with other poetry. Heidegger does not pursue this reasonable thought; instead, he immediately says that a dialogue between poetry and thinking (i.e., Heidegger's philosophical thinking) is possible—without claiming that it could be a "true" dialogue—and that such a dialogue offers something distinctive: "The dialogue of thinking with poetry aims to call forth the *nature* of language, so that mortals may learn again to live within language" (Heidegger 1971a: 161). So, already in the introduction to "Language in the Poem," Heidegger makes it clear that this essay is about mortals learning to live once again with language. There is no mention of our fellow animals.

Heidegger's essay "Language in the Poem" consists of a brief introduction and three parts. Two points are made in the introduction. I have already discussed one—the challenge of having a dialogue between the thinker and the poet. The other is that every great poet makes only a single statement, but a statement that remains unspoken in his/her poems. As developed throughout the essay, Heidegger says that Trakl's unspoken statement is about "apartness" as "the first beginning" (p. 194) of a moving away from the decay of human life. "The first beginning" evokes an identical phrase used in *Contributions*—along with its connection to "the other beginning"—which, Heidegger says, is only hinted at in "Language in the Poem." We can have confidence that "apartness" and "the first beginning" reflect Heidegger's own understanding, whether or not they are Trakl's. Furthermore, we can have further confidence that it is human life which concerns Heidegger in this essay, not that of our fellow animals.

In Parts I and II of the essay, Heidegger (1971a: 161) turns to Trakl's poetry. It starts, "Something strange is the soul on the earth." The "soul" is a reference to humankind, thus immediately telling us that it is the *human* who is at stake in this essay, and it is the human who is somehow seeking a place on earth. Shortly thereafter (Heidegger 1971a: 163), Trakl says of this human strangeness, "The thrush called to a strange thing [also translated "stranger"] to go under." In Heidegger's thinking, "going under" entails departing from the meaninglessness of this world with its focus on entities rather than being. It corresponds to the first level of profound boredom described in *FCM*; it is a moving away from inauthentic *Dasein*; it is the apartness that Heidegger identifies as Trakl's guiding thought. (The second moment of profound boredom—the moment of vision—remains concealed and futural to Trakl's poetry.) But what about the thrush? Is the thrush communicating with a particular human, i.e., the stranger? Is the thrush signaling something that is open to the human? Does it have a world that is shared with humans? That would indeed represent a change in Heidegger's thinking about animals. Of course, this is a direct quotation from Trakl, but still it is one that Heidegger chose to quote. What is clear is that Trakl is not providing his reader with a lesson in natural history, and it therefore seems unlikely that Heidegger is either, especially when thought as leading to a moment of vision. Further, at times, Heidegger talks about "mortals," and this signals that he is thinking ontologically. But still we might ask whether the thrush calls in the sense of inviting the human to go under or, alternatively, in the sense of repelling the human, and thereby "encouraging" him/her to leave this earthly place? Perhaps the thrush, with its inability to know being, merely makes sounds. Aware of this and its meaninglessness, the human finally sees his/her own connection to this and is repelled by the thrush. This interpretation is similar to the one Heidegger presented for *FCM*. Can we conclude this from this essay? No. But can we conclude that the thrush is open to the human and a world in Heidegger's terms? No. In fact, immediately after the quote, Heidegger simply says, "The soul is called to go under" (Heidegger 1971a: 164) without any mention of the thrush. And outside this one statement from Trakl's poem, Heidegger is singularly uninterested in the thrush. Perhaps Heidegger is simply letting Trakl have his own voice and is more concerned with the "going under" part than who it is that "makes" the human aware of the possibility of going under.

Another animal motif that features prominently is "the blue wild game" (also translated "blue deer"). Heidegger (1971a: 164–5) quotes Trakl, "The footfall of the stranger. Would that the blue wild game were to recall his paths …," and then asks, "In what sense is the 'blue wild game' to recall what is going under?" Heidegger here interprets "blueness" as the holy. Heidegger (1971a: 166–7) then says,

> Who is this blue wild game to whom the poet calls out that it recall the stranger? Is it an animal? No doubt. Is it just an animal? No. … This animal—the thinking animal, *animal rationale*, man—is, as Nietzsche said, not yet determined.
> … This animal not yet determined in its nature is modern man.
> By the poetic name "blue wild game" Trakl evokes that human nature whose countenance, whose countering glance, is sighted by the night's blueness, as it is thinking of the stranger's footfalls and thus is illumined by the holy. The name "blue game" names mortals [my underlining] who would think of the stranger and wander with him to the native home of human being.
> Who are they that begin such a journey? Presumably they are a few ….

So, the blue wild game is a human after all, not the human who "sees" historical being, but the one who, like the modern human being, is burdened by entities. With the help of the stranger, he/she now starts to move away from them, and in so doing, starts moving away from metaphysical/modern man. But in this transition, what Heidegger calls the "in-between," the human is not yet truly mortal, only "indefinite." Like *Da-sein* relative to *Dasein*, few humans are open to "going under." Thus, Heidegger is writing about the transformation of the human being. "The blue wild game" is Trakl's poetic image of the human in its "decaying" state, an image that Heidegger leaves intact, without discussing anything whatsoever about our fellow animals. In leading up to his conclusion, Heidegger identifies Trakl's other poetic images of the blue wild game in transition toward the in-between—shy game, gentle animal, animal face.

We still might wonder why Heidegger says of the blue wild game, "Is it an animal? No doubt. Is it just an animal? No." Is Heidegger here saying that humans are animals? Not at all. Heidegger is here giving Trakl his own poetic freedom to express his theme of apartness using animal imagery. Heidegger says, "The blue game is an animal whose animality presumably does not consist in its animal nature." So this "animal" does not have the essence of animality but that of the human.

Other images of our fellow animals also point toward that decaying state which the human is leaving behind in going under. "Blackbird's plaint falls silent" (p. 169) as the wildness is drawn toward total darkness and the silent blue holiness of the in-between. Likewise, "Fish and game soon glide away" (Heidegger 1971a: 170, 173) as the decaying world is left behind. And, "In his grave the white magician plays with his snakes," indicating that the decay (i.e., the malice of the snakes) is no longer affecting the human. As a last example, Heidegger (1971a: 186) quotes Trakl, saying, "The fervid lament of the animals," identifying those who are left behind. The "apartness" that guides Trakl's poetry, according to Heidegger, is another way in which Heidegger evokes the abyss separating humans from our fellow animals.

In Part III of the essay, Heidegger (1971a: 195) further develops Trakl's unspoken theme of apartness, focusing on a particular phrase—"*One* generation"—that he thinks "silently sounds the mystery" of Trakl's poetry. "One" points toward a hidden and unifying force in the blueness, the holiness. As for "generation," Heidegger says, "For one thing, it names the historical generation of man, mankind as distinct from all other living beings (plants and animals)." Thus, despite the variety of animal images, this essay is decidedly about humankind, not our fellow animals. There is nothing in it that should lead us to think that Heidegger has significantly revised, or even engaged, his prior understanding of animality. One way of corroborating this conclusion would be to look to his future comments. If they support earlier interpretations—primarily that presented in *FCM*—then that would support the idea that he had not changed his views in this essay, and that any such interpretations are simply fanciful. There is not a lot more that Heidegger has to say about our fellow animals by this time in his life, supporting the idea that the issue necessitated no further clarification, but there are hints that he has not changed his views (see below).

"Basic Principles of Thinking: Freiburg Lectures 1957"

(Essay of 1957; Heidegger 2012a)

Quoting from a poem by Hölderlin, Heidegger (2012a: 107) says,

> Let us note it well: it is the mortals who reach sooner into the abyss, therefore they are the ones who dwell in the refuge of death and are thus able to die. An animal cannot die; it comes to an end. This may go along with the fact that the animal cannot think. Thinking lives by an elective affinity with death.

Here Heidegger draws a connection between death—as a way that human *Dasein* lives—and thinking (*Denken*). By the time of this essay, thinking is no longer "metaphysical thinking," that is, the saying of beings. Rather, Heidegger's post-*Being and Time* usage of "thinking" has incorporated an abundance of distinct phrases and terms designed to capture different aspects, or perhaps different levels, of the saying of being. These include: being-historical thinking (*seingeschichtliches Denken*), inceptual thinking (*anfängliches Denken*), commemorative thinking (*Andenken)*, mindfulness (*Besinnung)*, knowing-awareness (*Wissen*), enowned thinking (*Ereignisdenken*), and innovative thinking (*Er-denken*). Perhaps "being-historical thinking" is his most all-encompassing phrase, and—given all that I've written about Heidegger—is likely outside the domain of our fellow animals, but most humans also. However, Heidegger mentions two other, related forms of thinking that are not so clearly outside the domain of animals: calculative thinking (*rechnendes Denken*) and thinking as representation (*Vorstellung*), but whether Heidegger would *really* consider these "thinking" is not clear. After all, in a 1951 essay Heidegger (2008: 373) provocatively says, "Science does not think," although here he meant that scientists do not directly and explicitly think being itself.

Later in the 1957 essay, Heidegger (2012a: 113) says,

> If we understand thinking as the privilege of the human, then the *belonging-together* of the human and being stands as our topic. Yet we see ourselves instantly compelled by the question: What does being mean? Who or what is the human?

Thirty years after *Being and Time* was published, Heidegger is asking the same question: What does being mean? And this question remains firmly attached to the human with little hint that our fellow animals need to be considered beyond what he provided in *FCM*.

"The Nature of Language"

(Essay of 1957–1958; Heidegger 1971b)

"Saying" is a special word for Heidegger that illuminates his understanding of language. It can refer to a cultural tradition that is both oral (e.g., proverbs) and performative (e.g., festivals). By combining thinking and poetry in "Language in the Poem," Heidegger (1971a) uses this understanding to address being. But saying can also be silent for Heidegger, where hearing precedes speaking. In this essay, Heidegger (1971b: 107) says,

> Language, then, is not a mere human faculty. Its character belongs to the very character of the movement of the face-to-face encounter of the world's four regions. … Language, Saying of the world's fourfold, is no longer only such that we speaking human beings are related to it in the sense of a nexus existing between man and language. Language is, as world-moving Saying, the relation of all relations. … Reserving itself in this way, as Saying of the world's fourfold,

> language concerns us, us who as mortals belong within this fourfold world, us who can speak only as we respond to language. … Mortals are they who can experience death as death. Animals cannot do so. But animals cannot speak either.

While it is interesting that Heidegger's understanding of language has changed since writing *Being and Time*, I remain focused on our fellow animals. Heidegger remains clear that the word "mortals" refers only to humans, not animals. Mortals die, not animals. Mortals can dwell in the fourfold in the sense that they are on earth, under the sky, and await the divinities. Animals have uncertain placement in the fourfold, and then only in the presence of mortals. Finally, animals cannot speak. Yes, Heidegger's thinking changes over the decades, but not in all respects. His human/animal dichotomy remains firmly intact.

"Time and Being"

(Lecture of 1962; Heidegger 1972b)

This lecture revisits Heidegger's understanding of being and temporality as they were presented in *Being and Time* some thirty-five years earlier, but now within the frame of the event of appropriation/adaptation (*Ereignis*). Appropriation recognizes the uniqueness of be-ing, contra a metaphysical understanding of the universality of the human essence. While he makes no direct mention of our fellow animals, Heidegger does suggest in "Time and Being" that being "reaches" only humankind and remains "concealed" for other beings (p. 12). But does being reach all humankind, or just a select few who are open to it? If the latter, as much of Heidegger's later oeuvre suggests, then Heidegger's understanding of "man" is likely similar to that of *Da-sein* and not to *Dasein*, let alone to the naturalistic entity we commonly recognize. Such an understanding might suggest that Heidegger is both anthropocentric and elitist (based on generating a cleavage within human *Dasein*; see Polt 2006 and Belgrano 2021), but since he is speaking ontologically, this suggestion, which is hinged in political (ontical) thinking, may not apply. Interestingly, Agamben (2004) uses Heidegger's cleavage within *Dasein* as a wedge to form a connection between our fellow animals and the large remainder of humanity (i.e., human *Dasein*), but it is clear that Heidegger would not concur.

Zollikon Seminars. Protocols—Conversations—Letters

(Lectures and letters from 1947 to 1972; Heidegger 2001)

In this essay, Heidegger (2001: 90) claims that the "as" relationship is necessary for the human to speak, and that its withholding leaves our fellow animals with nothing to "say." The frequency with which Heidegger has repeated these claims over the decades indicates their importance to him. Heidegger (2001: 286) repeats one other central point about the dichotomy between humans and our fellow animals, namely, that only humans have existence as their way of being.

> But in saying this, it is in no way asserted that material nature and organisms (plants, animals) are not real, are not beings, but are mere appearance. It is merely said that the reality of the aforementioned nonhuman domains is other than existence.

In this quote, Heidegger refers to our fellow animals as "organisms," but with a usage that disavows the common biological usage. In *FCM* (p. 235), Heidegger says,

> The term "organism" therefore is no longer a name for this or that being at all, but rather designates a particular and fundamental manner of being. We can briefly characterize this specific manner of being as a proper peculiarity with the capability to create organs.

"Proper peculiarity"—originally used in *FCM*—continues to refer to the organism's intrinsic absorption in itself (i.e., its way of being as captivation), and it remains inapplicable to the human, whose manner of being is not one of absorption but of an ecstatic "looking" away from itself, or existence.

Now a brief comment on embodiment, which is a topic little addressed by Heidegger (see Schalow 2006). First of all, there is no reason to doubt Heidegger's acceptance that all animals, including humans, are embodied. But, what is the role of the body in animal life when considered ontologically? Fortunately, *Zollikon Seminars*—based on letters and lectures given to medical doctors—is one place where the issue of embodiment is briefly, but directly, addressed. After considering what he says in *Zollikon Seminars* about human *Dasein*, I suggest that certain problems remain as regards our fellow animals.

Heidegger (2001: 86–7) says,

> the body is not a thing, nor is it corporeal thing, but each body, that is, the body as body, is in each case my body. The *bodying forth of the body* is determined by the way of my being. The bodying forth of the body, therefore, is a way of Dasein's being.

So, the body is first and foremost dynamic in its bodying forth as *my* body. But there is also an abiding aspect of the body that can become manifest. In contrast to bodying forth, the corporeal thing has its being outside of *Dasein*'s existence and belongs to no one in particular. He continues,

> bodying forth is co-determined by my being human in the sense of the ecstatic sojourn amidst the beings in the clearing. … The limit of bodying forth changes

> constantly through the change in the reach of my sojourn. In contrast, the limit of the corporeal thing usually does not change.

While the language of "ecstatic sojourn" and "clearing" are post-*FCM*, they are hinted at in his usage of *Da-sein* versus *Dasein* and *ek-sistence* versus existence. Bodying forth is *Da-sein*'s way of being, and that way of being includes "saying" and "speaking," in which bodying forth participates through hearing and seeing. Not that it is the body (or the ears or eyes) which hears and sees but rather that it is human *Da-sein* ontologically tethered, as it is, to that body. By contrast, a corporeal thing does not sojourn and usually is of a constant form. Similar to an understanding of "world," "body" has both a dynamic aspect (bodying forth) and an abiding aspect (the body, but still not to be viewed as corporeal), but it is clear that the body can only arise in the presence of bodying forth, and thus the dynamic aspect is primordial.

In experiencing the bodily, Heidegger says, one must begin with the human being as existing. In being open and relating to the world, the human being responds to it through his/her comportment. Absent a bodily nature, this would not be possible. All aspects of our bodily nature, down to the most microscopic levels, belong in their essence to existing, which is the human being's particular way of being. This way of being cannot be objectified because *Dasein* is always already being-in-the-world, and its manner of existence cannot be derived from anything else. Any objectification which does occur (and it oftentimes does!) is not, therefore, originary in Heidegger's view, and if it claims to be so, it is simply misguided, or so Heidegger thinks. Heidegger provides the example of the ability to direct our attention (e.g., in front of us, behind us, to the side), something that our existentiell bodily being is capable of; whereas, any such ability of a corporeal body considered as a present-at-hand object is simply not understandable, even with detailed mechanistic descriptions that biologists can offer. Heidegger (2001: 234) concludes:

> But the decisive point in our context is our insight into the immediate emergence of all of our so-called material, bodily nature from the physically intangible capacities for receiving-perceiving and for comporting oneself, in which our Da-sein in its unfolding essence consists.

So much for the body and human *Dasein*. What about the body and our fellow animals? Heidegger is clear that "bodying forth" is a feature of human *Dasein*, and not of our fellow animals. Are our fellow animals simply corporeal things or something else? Heidegger does not expand on this. Once again, we have a linguistic demarcation between the human being and our fellow animal which presupposes that very demarcation. Given the close relationship between bodying forth and comportment, this may not be surprising. We are left with Heidegger's unsettling thought that our fellow animals are embodied in a different sense from that of humans, and that captivation must be central to any understanding of our fellow animals.

Heraclitus Seminar

(Seminar of 1966–7; Heidegger and Fink 1993)

This book records the dialogue among Heidegger, his colleague Eugen Fink, and students who are trying to understand the fragmentary writings of Heraclitus. A few interesting comments concerning our fellow animals are made, but they are all attributed to Fink with no further commentary from Heidegger.

"Seminar in Le Thor 1969"

(Seminar of 1969; Heidegger 2003b)

In this seminar, Heidegger makes an interesting claim about speaking. He says that speaking is a feature of world(ing)—of the fourfold (earth and sky, divinities and mortals)—and cannot be understood as being confined to humans the way many naturalists presume. Even so, as "the mortals" are one moment in a *unity* of four moments, there remains a necessary linkage between speaking and human *Dasein*. Of course, saying that speaking is a feature of the world still does not mean that our fellow animals can speak. To speak requires an understanding of being, and that is something that is denied to our fellow animals. Speaking is worldly, but as the shepherd of being—the one who provides shelter for being through language (Heidegger 2003b: 63)—human *Dasein* still has a unique role.

....

My aim in this chapter (and the previous one) has been to critique Heidegger's view of our fellow animals vis-à-vis that of human *Dasein*, and to a lesser extent that of nonliving entities. Following Heidegger, I have refrained from making strong distinctions among the many species of our fellow animals. I have argued that Heidegger's view (ontologically considered) does not majorly change after the *FCM* lectures of 1929–30 and that the account in *FCM* is compatible with his less developed account in *Being and Time* published in 1927.

Prior to 1927, Heidegger's thinking about our fellow animals can be thought of as "gestational." For example, in 1926—one year before *Being and Time* appeared—Heidegger (2007: 283) says,

> Even a jellyfish already has, when it is, its world. Something like a world, a being that it itself is not, is revealed to it, uncovered. For the chair, however, world is beyond any possibility of being discovered or concealed.

Here, the jellyfish has "something like a world" rather than a world itself, but Heidegger clearly remains receptive to von Uexküll's ecological way of thinking. But in 1927, he enunciates the ontological difference (Heidegger 1962,

1988), and the implications for "world" become clearer to him, although the transitional nature of this work is still manifest in his explication of four distinct understandings of "world." By the time of *Being and Time*, Heidegger has come to think that von Uexküll's ontical understanding depends on a deeper ontological *mis*understanding, namely, that the world and worlding of humans and our fellow animals are not distinguishable in kind.

Being remains central to Heidegger throughout his career. If I am going to challenge Heidegger's understanding of our fellow animals, it must be through a revised understanding of their way of being. To that I now turn.

Chapter 3

Expanding Heidegger's Legacy: Reinterpreting an Animal's Way of Being in the Light of Evolutionary Thinking

In this and the following two chapters, I continue to think in Heidegger's ontological framework by using his concepts—*Dasein*, abyssal thinking, existence, captivation, way of being, temporality, spatiality—but now I place greater emphasis on interpreting the ontical record as found specifically in evolutionary biology and ethology. In this chapter, I argue in broad outline that evolutionary thinking can add a historical dimension to comparative ethological studies, converting them into evolutionary ethology when past behavior is inferred from analysis of extant organisms (see Gunasekaram et al. 2024 for a recent example) and into ethological paleontology when past behavior is inferred directly from extinct organisms (e.g., their fossils). Doing so opens up the possibility of an evolutionary analysis of *Dasein* itself, or so I claim. Relatedly, Heidegger says, "the ontical is only *one* domain of entities. The idea of Being embraces both the 'ontical' and the 'Historical'" (*Being and Time*: 455). While I consider my endeavor to be a justifiable extension of Heidegger's own thinking, I have no doubt that Heidegger himself would have seriously questioned this manner of proceeding.

But before beginning my commentary on evolutionary thinking, I want to acknowledge that numerous other authors have commented—largely outside of an evolutionary framework—on Heidegger's view of our fellow animals and of the abyss separating them from humans. Indeed, in an early version of this book, I devoted an entire chapter to their thinking in relation to my own. The vast majority are trained philosophers, unlike myself. Some focus solely on their understanding of Heidegger's views, and voice support or criticism thereof. Others outline their own understandings, for example, by prioritizing a particular ethical framework, but I am not aware of anyone else who engages our fellow animals as the individuals they are through a deeper understanding of evolution and ethology without also turning away from Heidegger's thinking. Eleven authors who explicitly respond to Heidegger's own understanding of our fellow animals are: Bruce V. Foltz (1993, 1995), William McNeill (1999), Matthew Calarco (2004, 2008, 2015, 2024), Giorgio Agamben (2004), Tracy Colony (2007), Stuart Elden (2006), Brett Buchanan (2008), Andrew J. Mitchell (2011, 2015), Chad Engelland (2015), Beth Cykowski (2021), and César Alberto Pineda Saldaña (2025). I learned much from reading them. Interesting, but more tangential to my project, are other authors who primarily use Heidegger's writings as a reference to develop their own views (Gadamer 1994; Okrent 2007; parts of Foltz 1995; Derrida 2008; Chamberlain 2013; Storey 2015).

I wish to draw particular attention to a seminal work by Cary Wolfe (2013), namely, *Before the Law: Humans and Other Animals in a Biopolitical Frame* (see particularly: Chapter VII), which, like my own accounting, claims that (some of) our fellow animals can also be in *Dasein*. Wolfe's critical account of *Dasein* is brief—and briefer still for *Dasein*'s being as temporality and spatiality. Unlike my own account which relies heavily on interpreting the findings of biologists, Wolfe presents his views mostly in conversation with philosophers and other scholars of comparative literature (particularly: Jacque Derrida, but also Michel Foucault, Roberto Esposito, Richard Beardsworth, Martin Hägglund, Matthew Calarco). But while Wolfe rejects Heidegger's abyss between humans and animals, he also rejects a symmetrical universalism of all life; instead, he sees—along with Derrida—many nonabsolute but noteworthy divisions within life based on an ethic that he calls "biopolitical." In this ethic, the importance of ethology and the study of animal diversity are noted, but the biological content he presents is minimal.

In writing this book, I have chosen to avoid commenting on Heidegger's thinking as it might relate to modern environmentalism, including Deep Ecology (e.g., Zimmerman 1993, 2003; Foltz 2014). This is because Heidegger's thinking remains focused on being and on the being of individual human *Dasein*, even when talking about "a people" (e.g., see *Contributions*: 316). Extending his ontological thinking from specific ontical claims about individuals and their groups to the environment, broadly conceived, is beyond the scope of my book.

As I start down my own path, I wish to repeat that I have not chosen Heidegger simply as a foil for my own thoughts; rather, I find his thinking to be insightful and worth pondering, and I want to acknowledge upfront that by reading him, my own view of humans and our fellow animals has changed. But, as the biologist I am, I want to know the basis of his claim that there is an ontological abyss separating our fellow animals and humans, even though evolution provides one form of compelling evidence that there is no ontical abyss. Thus, I return to the question of the relationship between the ontological and the ontical. What does Heidegger mean when he claims there to be an ontological abyss separating our fellow animals and humans? The abyss that Heidegger is talking about is in being itself, and there is no deeper ground (even if ungrounded) than this. Of course, we do not have to accept this claim. But what does Heidegger actually say about the relationship between the ontological (that is, being) and the ontical (that is, beings)? Near the very beginning of *Being and Time*, Heidegger says,

> In so far as Being constitutes what is asked about, and "Being" means the Being of entities, then entities themselves turn out to be *what is interrogated*. These are, so to speak, questioned as regards their Being. But if the characteristics of their Being can be yielded without falsification, then these entities must, on their part, have become accessible as they are in themselves. When we come to what

> is to be interrogated, the question of Being requires that the right way of access to entities shall have been obtained and secured in advance.
>
> (§3: "The Formal Structure of the Question of Being," p. 26)

So, for Heidegger that "right way of access to entities" turns out to be through a deeper understanding of being. About this, I am in agreement. But, Richard Polt (2006) argues that Heidegger's philosophy is underdeveloped with respect to how his understanding of human *Dasein*'s being actually relates to the ontical and, in particular, to the human as an embodied being amidst other embodied beings. About this, I am also in agreement. Now consider this: If (some of?) our fellow animals can also be in *Dasein*—as I shortly claim—then Polt's argument applies to them as well. It is from out of the ontical contexts provided by evolutionary (this chapter) and ethological (Chapters 4 and 5) thinking that I strive to rethink Heidegger's abyssal claim—not as a source of determinative evidence but as a basis for making a different ontological interpretation, one that is more consonant with what it means for *Dasein* to be in a world.

Section A. Evolutionary Thinking and Way of Being

How can we say that humans and our fellow animals are or are not all animals unless we already know their way(s) of being? That is Heidegger's point. But there is an additional point: Heidegger's understanding of "animal" in *FCM* is not that of biologists. In particular, Heidegger views animals neither through the rational/irrational divide (as psychologists might) nor as machines (as functional biologists might) nor as objects (as morphologists do). Instead of talking about what animals *are*, he talks about the differing ways of being of our fellow animals and humans. In doing this, he sometimes relies on their behavior (or "comportment" when referring to humans) in support of his ontological claims, since behavior can be seen as a holistic feature of the particular being. In this sense, "*way* of being" is an ontological interpretation of an ontical understanding. But Heidegger is particularly concerned that any behavioral (ontical) features explicitly acknowledge his own pre-understanding of their abyssally differing being: captivation for our fellow animals, existence (and freedom) for humans. In other words, "way of *being*" is based on a pre-understanding of being itself. But where did this pre-understanding of being come from if not based on some interpretation of the ontical? Does Heidegger think that his understanding of being is even sensitive to the ontical, that the ontical (e.g., biology, "good" poetry, philosophical systematizing) must necessarily support his view? Despite his emphasis on questioning rather than providing definitive answers, Heidegger seems to want to lead us in the ontological-first direction. But then, in *FCM* he raises the specter of his own anthropocentrism. It is my claim that if we think about biology as holistically as possible while still insisting on making definite observations, then we can have a pathway, albeit imperfect, to an interpretation of the ontological. Heidegger and I already agree that observations by ethologists are not to be dismissed, even if we bring to them different interpretations, both ontical and ontological. But additionally and despite Heidegger's disclaimer, I claim that some observations by evolutionary biologists sufficiently approximate Heidegger's holism criterion, plus they add an important historical dimension.

To address this, I believe we should turn to what Heidegger himself[1] was largely uninformed about, namely, Darwinian evolution,[2] but without ignoring an important complexification.[3] As Darwin explained using multiple approaches and examples, the case for biological evolution rests on its making improved sense of an ontical world. That is why Darwinian evolution is widely accepted today, with natural selection providing a central, but not only, "mechanism" for evolutionary diversification. But "improved sense" here is predominantly an ontical claim, and it entails making statistical inferences, not arriving at absolute claims. Relatedly, evolutionary thinking cannot "drive" ontological thinking; that would invert Heidegger's way. But we do need, I believe, better connection to the ontical than Heidegger's own analysis of our fellow animals provides in order to make sense of that ontical world that he describes in *Being and Time* and particularly in *FCM*. By my way of thinking, Heidegger has not established that evolutionary theory entails a misunderstanding of being. Supporting this claim is one of the aims of this chapter.

Evolutionary (comparative) studies of behavior have burgeoned since Heidegger's time, and my question is: How do we think ontologically about these? Heidegger's own ontical thinking about our fellow animals was broad-brush. He was not a biologist and he relied on a limited number of biologists and a highly limited number of studies for his ontological interpretations. In making his interpretations, he was much more influenced by philosophers—Plato, Aristotle, Kant, and Nietzsche most prominently—than by theoretically minded biologists. This bias is questionable. For example, from the time of *Being and Time* onward, he dismisses von Uexküll's talk of an animal's world as ontologically misguided, and in *FCM* he reinterprets their world as an encircling ring of drives. What ontical awareness gave rise to this reinterpretation of all animals, most of whom Heidegger knew nothing about? The necessary reciprocity between the ontical and ontological—between being (way of being) and beings—indicates that we cannot think about that distinction as identifying a rigid boundary, even if we sometimes talk about it in that manner. So, for example, when presented with new ontical observations ("data"), we should not be closed to the possibility of new ontological thinking that challenges such data as data.

Heidegger's thinking about *Dasein* as found in *Being and Time* and *FCM* is not diachronic; indeed, it borders on the universalist. Diachronic thinking, which includes evolutionary thinking and developmental thinking, was not of great interest to Heidegger, except to question its ontological significance. In my opinion (and that of others), the consequences for Heidegger are severe, even though there is richness in Heidegger's more "synchronic" account.

Likewise, Heidegger's interpretations (articulations) of *Dasein*'s way of being are not scientific arguments, but neither will mine be. A scientific approach pretends to look at being from the outside, and for Heidegger that makes no sense. Instead, we always find ourselves already in (and "part of") a world we are trying to interpret—early Heidegger views being as a ground but importantly as an ungrounded ground. Later, he comes to consider talk of ground as too easily embedded in metaphysical thinking, and so his language changes; for example, "the open" takes on some aspects of ground although without a strong sense of abiding, but even here, he still distinguishes the open from (what has been translated as) ab-ground. Such an abstract approach may not satisfy many biologists, who rarely talk about being, but Heidegger would counter that even scientific statements are interpretations, albeit typically without acknowledgment of their underlying presumptions about being. Those who deny that being can be conceptually separable from beings simply cannot be open to Heidegger's thinking. For Heidegger, addressing being and justifying one's views of it operate at a different "level" than mere correspondence between statements about entities and those entities themselves, as scientific "correctness" is oftentimes understood (e.g., as expressed in the laws of classical physics). But in this regard, the historical sciences are not so much like classical physics. Evolutionary thinking aims to make statistical sense of the way the world operates in its contingency rather than claiming a deterministically correct understanding of the way it *is* (Brandom 2011). Making sense of the world and saying, as Heidegger does, that truth lies in (way of) being itself are not unrelated, or so I claim.

Since *FCM* was written, and particularly starting in the latter half of the twentieth century, there have been spectacular, though not indefeasible, advances in the phylogenetic reconstruction of animals based on an analysis of their ontical characters, such that Heidegger's claim about the necessity of a prior, non-objective understanding of essences now seems outmoded, even quaint, to biologists. Studying morphological—and more recently molecular and behavioral—character traits of various species, biologists have built an increasingly unified "tree" of life, including a convincing demonstration that the human lineage diverged from the chimp/bonobo lineage approximately six to eight million years ago. From an evolutionary perspective, we might ask: Are Heidegger's existentialia to be found on one, and only one, extant terminal twig of the tree of life—that is, our own—rather than being more broadly distributed? Do existentialia appear all at once on that twig, and when? Must existentialia come as the full "package" offered by Heidegger for human *Dasein*, or might there be species-specific differences? Heidegger does not address these questions in *FCM*. If he had, he would have particularly considered our sistergroup, the chimps and bonobos. Instead, he intentionally avoids discussing animals that bear striking similarities to humans.

But wait! Am I here equating existentialia (for humans) + categories / categorials (for the other animals) with biological characters (for animals, including humans), counter to what Heidegger argues? And the answer is: This would not be a good description of what I am doing. Heidegger's distinction is not arbitrary in that existentialia and categorials—unlike biological characters—are of a higher order and fundamentally non-objectivist (see List 1.1). But the further claim that existentialia are to be found only in the human lineage still requires more than simple assertion; it requires a comparative analysis that takes into account our deepest understanding of animals in their ontical diversity, a diversity that Heidegger barely explored. Following Heidegger, I agree that behavior (or comportment in humans) offers a particularly rich source of biological characters for ontological interpretation. Here, I am not saying that biological characters can be transformed in a one-to-one manner into and from existentialia and categorials. Their ontological understanding always requires an interpretation that is beyond so-called direct observation.[4]

To understand what Heidegger means by "categories," we must acknowledge his pre-understanding of the abyssal distinction between the animal and human ways of being—namely, captivation and existence. In *FCM* (PART TWO, end of Chapter Four), Heidegger articulates captivation through six structural moments: withholding, being taken, absorption, openness for something else, the structure of encirclement thus given, and captivation as the condition of the possibility of any kind of behavior. This, then, is what he offers biologists when they think about the animal with its characters. Similarly, in *Being and Time* Heidegger articulates human *Dasein*'s existence through his discussion of "being · in · (the) world," and he further articulates it through existentialia (see List 1.1). This is what he offers those in the human sciences when they think about human characters. But must we accept Heidegger's abyss? I think not, and I think the ethological studies presented in Chapters 4 and 5 provide the clearest starting point for an argument as to why.

But note further: In terms of making an ontological interpretation, it should not matter whether behavioral characters are inferred from living animals (as ethologists do), their higher-level taxa (as evolutionary ethologists do), or from remains of individual animals whose higher-level taxa (e.g., the species) may be extinct (as paleontologists do). These three scientific endeavors entail somewhat distinct methodologies and—in the order presented—increasing levels of interpretative difficulty, but evolutionary endeavors in particular open up the additional possibility of understanding historical changes in *Dasein*.

Heidegger's broad ontological claim about all animals cannot dismiss the challenge presented by evolutionary thinking. From the fossil record, we know that functionally important character traits have appeared gradually—rather than all at once—and that such changes have been documented even within the human lineage. We know this, in part, because paleontologists have discovered remains of multiple, distinct, and now-extinct human species, that is, species within the genus *Homo*. We might wonder whether individuals of these other human species (e.g., *Homo neanderthalensis*, whose remains were first discovered sixty years prior to the *FCM* lectures) could be in *Dasein* too, or whether Heidegger would consider them and their way of being to be "animal." And, what about early forms of our own species (i.e., premodern *Homo sapiens*)?

By choosing to ignore the tree of life and its fossil record, Heidegger ignores the evolutionary past, even though he pays close attention to a highly restricted portion of the historical past, that is, that of philosophers in Ancient Greece. In so doing, we might ask whether Heidegger is unduly prioritizing access to a particular, ancient written record over access to the fossil record. I think so. Both forms of record can be studied as objects and interpreted in terms of a way of being, even if paleontologists, for example, seldom go in this direction. Scientific objectivity should not be understood too narrowly. For example, the evolutionary biologist and ornithologist Richard Prum has (controversially) argued (i.e., interpreted) that some female birds—among other animals—choose mates based on an aesthetic criterion, namely, the beauty of male feather patterns (summarized in Prum 2017; see Prum 2018). Of course, fossils and the ancient written record may not be equally contentful, but that provides no reason for ignoring the fossil record. Thinking about fossils—human and otherwise—could have provided an opening for Heidegger to revise his ontological understanding of the evolutionary past, but doing so would have undercut his claim that studying objects as scientists do is ontologically misguided.

Given that evolutionary novelty emerges from (already existing) ancestral states, evolutionists oftentimes speak about character "transformation," reserving the word "novelty" for particularly large quantitative transformations in some feature, or for describing a qualitative feature that appears de novo at a higher level of organization (e.g., a heterochronic, heterotopic, or heteroposic change in developmental expression leading to a morphological novelty—Gould 1977, 1999, 2002; Hatzopoulos and Regier 1987). Had Heidegger been interested as a philosopher in human fossil remains as such, he might have presented a more nuanced understanding of "essence" that acknowledges its genesis over evolution. While the later Heidegger does come to acknowledge that being itself undergoes epochal (historical) changes, he describes such changes as inexplicable, as though the other gradual (ontical) changes that surround changes in being itself cannot even provide a meaningful context, let alone diminish the mysteriousness of such changes in being. As examples, we could ask: Might not the emergence of our ancestors from an arboreal existence to a bipedal and highly mobile one meaningfully contextualize a change in the way of being? Might not our increasing capacity for comporting in the world contextualize a change in our way of being? Might not our increasing socialization as a species contextualize a change in our way of being? In *FCM*, Heidegger does not deny the utility of some biological explanations, although he unfortunately does dismiss evolutionary ones. But his principal concern seems to be that any biological explanation—with its talk of causes—should not be seen to have a determinative priority over the interpretative task of philosophy. On this, I agree. But Heidegger is hesitant to give *any* role to the ontical when interpreting a way of being, even when ethology comes into play. It is a position that no longer makes sense.

In *FCM*, Heidegger claims that evolutionists are necessarily opposed to holistic thinking. But what about the many modern natural historians who observe whole-animal behavior? I understand them, broadly considered, to be evolutionists too. Should their ontical observations be dismissed because their understanding of an animal's way of being may, or may not be, Heidegger's? Heidegger did consider the thinking of ecologist Jakob von Uexküll to be ontologically misguided, but still, he accepted von Uexküll's scientific (ontical) observations largely without modification, and he then went on to reinterpret them ontologically. One of the "misguided" ontological presumptions of von Uexküll was to treat the various creatures he studied as objects. Unfortunately, Heidegger never addresses how else von Uexküll the biologist might have observed other creatures. And regardless, Heidegger was still quite willing to use, and interpret, von Uexküll's observations, so somehow, they remained accessible to ontological thinking that was not misguided.

But are even clearly reductive evolutionists really necessarily opposed to holistic thinking? The evolutionary geneticist Richard Lewontin (2000, 2001) points out that while biological evolution is a holistic "process," not all aspects are deeply intertwined, and the role of biologists is to identify those articulations (i.e., those aspects of the whole that are conceptually separable) by using, for example, Darwin's theory of natural selection. Such articulations are scientific interpretations that must remain open to revision specifically because evolution is fundamentally holistic (and statistical), and the meaningful articulations cannot be known ahead of time or with certainty. Although he would not like the comparison, Heidegger seems to be doing something similar with regards to his holistic interpretation of being. He talks about being's (and way-of-being's) internal "structure" by identifying moments, or aspects, that illuminate the whole. These moments connect to the ontical, albeit in tension with the ontological, for example, the connection between mortals in the fourfold and the ontical human. These moments are neither articulated processes like functional biologists study nor suites of articulated characters like evolutionary biologists study; rather, they are hints given to finite beings like us of being itself.

So, does Heidegger's claim ring true for evolutionary biologists in particular—that they have a warped view of being which leads to ontical misinterpretation? Heidegger is correct that evolutionists (in this case, evolutionary ethologists and paleontologists) treat animals as objects who "have" various characters, and that these characters are not interpreted in the same manner as the existentialia of human *Dasein*. But how could it be otherwise, given the barriers of language and life-history that separate us from our fellow animals, although less so for our own pets? And equally poignantly (at least for me), how else would Heidegger propose that paleontologists study the fossil record (i.e., the fossils themselves plus the adjoining clues about their environs)? Are examinations of the fossil record based on a misguided understanding of being? Do they have no significance for an understanding of being? Or, do they instead *enable*—or, at least, work toward enabling—an interpretation of being in as direct a manner as possible? I favor this latter understanding. As an example, the fossil record has shown that certain modern birds—with their advanced behavioral capabilities which firmly place them within *Dasein* (see Chapter 5, Section D)—originated within the now-extinct dinosaurs, thereby providing a link to connect specific dinosaur/bird characters with putative behavior (see Brusatte 2017). This is an important step toward understanding what it would be like to be an archosaur, living or extinct. Fossils present paleontologists with technical challenges that ethologists who study living organisms do not have, but their task is neither hopeless nor misguided. If an understanding of a way of being is one's aim, scientific studies of behavior (or paleontology) in their objectivity should not be considered misguided, but only depauperate or partial. After all, in *Being and Time* Heidegger argues that an apophantic interpretation (typically in terms of one's writing) is a diminished form of hermeneutic interpretation, and yet, there he was, writing about being.

Paleontological studies open up the possibility of being able to interpret an extinct animal's way of being. This requires inferring behavior from fossils, along with their environmental context, and building a web of ontical relationships that together enable an ontological interpretation. Of course, the data are much sparser and less rich than what Heidegger brought to bear for interpreting human *Dasein*. But to even have a chance of understanding the being of extinct animals more fully, we must take an interest in them as they are presented to us, something that Heidegger refuses to do. To definitively lump all of our fellow animals into one way of being (captivation)—as Heidegger does—and one that is radically distinct from that of the human (existence), is not to take an interest in them; rather, it is to abandon them. Of course, it can be equally problematic to thoughtlessly leap to the other extreme by saying that existence as understood for human *Dasein* is also to be found in all of our fellow animals. This reverse-anthropocentrism is also problematic since humans remain the standard by which existence in our fellow animals is evaluated. Instead, we ought to remain responsive to the ontical record and humble about our ontological claims. That is what I aim to do in Chapters 3–5, by arguing that existence as a way of being is to be found in at least some of our fellow animals.

Importantly, later Heidegger says that the being of humans has changed over historical time. Sometimes he discusses this through seminal philosophers (e.g., Plato: being as *idea*; Aristotle: as *energeia*; Kant, as position; Hegel: as absolute concept; Nietzsche: as will to power—see *Time and Being*; Heidegger 1972b), other times through historical epochs (e.g., Ancient Greek, Roman, medieval, modern, present [age of technology]—see *Parmenides*; Heidegger 1992b), and still other times through being itself (e.g., being as presence of beings, as the Christian god's creativity, as objectivity [and subjectivity], as positionality/enframing—see "The origin of the work of art" [Heidegger 2002b] and "The age of the world picture" [Heidegger 2002a]). We might wonder whether the human's way of being remains existence even when being itself changes. Despite my own uncertainty over how Heidegger is using "being" and "way of being," I think he is clear that existence remains human *Dasein*'s way of being. But what is less clear is whether *Dasein*'s existentialia (see List 1.1) are fixed. If being itself can change, then why not also *Dasein*'s existentialia? And if they do change, then why should that change be confined to *historical* periods, and not also prehistorical/evolutionary periods? Comparative behavioral studies and studies of fossils, when placed in an evolutionary context, open up such possibilities for discovery.

So what might be some pivotal evolutionary transitions across all life forms? Smith and Szathmáry (2000) identify eight: (1) the transition from independently replicating macromolecules to their enclosure by a membrane (i.e., the origin of protocells); (2) the transition from independent replicators to their covalent linkage into chromosomes; (3) the transition from RNA as both enzyme and depository of information to proteins as the predominant form of enzymes and DNA as the sole depository of information; (4) the transition from cells without (membrane-bound) organelles to cells with such (e.g., nuclei, mitochondria and chloroplasts); (5) the transition from asexual reproduction to sexual reproduction; (6) the transition from single-cell organisms to multicellular organisms; (7) the transition from largely solitary individuals to coherent colonies that have internal differentiation; and (8) the transition from primate societies without language to human society with language. Each of these transitions is evocative of numerous possible accompanying behavioral changes of a fundamental (biologically speaking) sort. "Possible" is the key word here because these evolutionary transitions open up possibilities (not to be thought of as causes) for a creature's relating to its world. These possibilities are then explored and selected through (or "in") further evolution.

Of the eight transitions, #6 likely characterizes the *origin* of the animal lineage as it diverged from that of our nearest living relative, the unicellular (and colonial) choanoflagellates. But while multicellularity may be necessary for there to be existence as a way of being, it cannot be sufficient, since it is widely distributed outside animals and is even found in prokaryotes.

Transition #7 is worded such that it too can characterize organisms outside animals, including even prokaryotes. But eusociality, as the most highly structured form of sociality recognized by biologists, is confined to animals and has independently evolved numerous times—for example: in insects (some ants, bees, wasps, termites, aphids, thrips, and beetles); malacostracans (some shrimp); rodents (mole-rats); and humans, although not as strongly as in some others. Could the evolution of eusociality—along with less highly structured social forms found, for example, in primates, cetaceans, and birds—have enabled existence as a way of being to emerge? Would such a sociality hypothesis indicate that existence could have independently appeared numerous times? But is it clear that non-social organisms cannot also have existence as a way of being? And besides, are there any fully non-social animals? These questions beg for further ethological studies. The following two chapters are my (incomplete) responses to these questions.

Transition #8 is worded anthropocentrically, and it highlights both human social organization (already discussed as transition #7) and the human form of language. But if, contra Heidegger, we reorient from human language to communication skills, we find that they are broadly distributed across animals and rely on various senses (e.g., smell: ants [through pheromones] and dogs [through urine]; touch: bees [in following the waggle dance—see next chapter]; hearing: birds [through their song]; sight: octopuses [through their warning coloration]; taste: catfish [through their "swimming tongue," see Yong 2022]). In thinking about existence as a way of being, are we justified in considering language in this broader context? Could Heidegger have benefited from deeper engagement with ethology?

Is Heidegger's ontological abyss between humans and our fellow animals in conflict with the largely continuous nature of evolution? If the unique features of modern, fully functioning, adult humans are to provide a pre-determined standard, then, yes. But why should we predispose ourselves in this manner when we know so little about the rest of life, including our own ancestors? Here, I need to add that Heidegger's oft-repeated claim that language cannot be viewed as something added on to the human because there can be no human without language is correct but misleadingly so. It is correct in that a human without language—and for purposes of clarity, without any of the other forms of communication listed in the preceding paragraphs—would be fundamentally different from one with language. It is misleading in that it draws overly sharp borders between humans and other creatures with advanced communication skills, including our sister species—chimpanzees and bonobos. Relatedly, Heidegger avoids talking about other members of Hominina (i.e., all creatures within the human lineage after its separation from the chimp + bonobo lineage), such as Neanderthals, whose remains were discovered well before Heidegger was born. But even more provocative are more recently discovered hominina (e.g., *Australopithecus* species) who are much older than Neanderthals and whose language skills were likely to be much less than modern humans based on scientific analysis of the fossils. Heidegger is not necessarily denying that—biologically speaking—modern humans evolved from a language-less ancestor, although I am uncertain whether he accepted any objectivist evolutionary account as other than fundamentally misguided. But I do believe that his primary concern with the biological account of evolution and its gradualistic assumption is that it is presented by scientists as all-encompassing and sufficient for understanding human evolution. I think this is a warranted and reasonable concern. Nevertheless, it still seems that Heidegger fails to see that an event of (unexplainable) appropriation (e.g., awareness of a new "world") does not have to exclude there being enabling gradualistic changes that biologists can document and that can enrich our understanding of the event, and that can even diminish (some of) the "event's" mysteriousness. For example, as regards the counterfactual just mentioned (i.e., a language-less human), I agree with Heidegger that this makes little sense when language is thought as a two-state condition only. But language did not arise in a moment, and though there may well be scientifically unexplainable transitions, many (ontical) transitions can be—and have been—usefully documented in scientific terms. Whether this ontical analogy extends to the ontological—as Heidegger claims it cannot—should remain open.

Let us return to a consideration of major evolutionary changes, but now within animals. There are many candidates in addition to those listed by Smith and Szathmáry (2000). For example: (a) The evolution of bilateral symmetry early in animal evolution. Bilateral symmetry optimizes an organism's maneuverability (Holló and Novák 2012), providing it with a stronger sense of directionality, that is, where it is going. This could be considered as an early step in the evolution of selfhood. (b) The evolution of encephalization, that is, the increase and complexification of neural clustering into a brain (Narayan and Verma 2019). By regulating other cells/tissues/organs, the brain coordinates the activity of the entire organism, which is a central aspect of what it means to be an individual. (c) The evolution from an exclusively marine existence to an exclusively terrestrial existence. This occurred in the vertebrate lineage, but also multiple times with arthropod lineages—and protostomes generally. Interestingly, terrestrial ancestors of some vertebrate lineages have (secondarily) reverted to being marine organisms, for example, cetaceans such as whales. Such transitions are not simply a change in environment; major alterations in sensory perception are entailed. (d) The evolution of a rigid exoskeleton in the arthropod lineage. Such an adaptation signals the likely appearance of new identities, such as predator and prey. (e) The evolution of eusociality. This has occurred multiple times within animals, perhaps earliest in coral, but also quite early in Isoptera (termites) and Hymenoptera (wasps and bees). Sociality can be seen as a fundamental alteration in what it means to be an individual and a "people." (f) If we look within the modern human lineage (Hominina), numerous acquired abilities (not all of which are unique to this lineage nor are they to be thought of as biphasic) are likely to have had profound consequences. Some of these are: upright walking (bipedalism) with consequent changes in habitat and enhanced migratory capabilities (e.g., see DeSilva 2021; a popular and more general paleontological account of human evolution can be found in Shubin 2009); (ongoing) technological sophistication such as fire building, toolmaking with consequent advanced hunting and fishing techniques, agriculture, and urban living; and enhanced imagination, innovation, mimicking, cooperation, mind-reading, time-traveling, symbolic thinking, moral thinking, and abstract thinking. I am confident that if we knew other lineages as well as our own, similarly impressive lists could be constructed for them.

If we think deeply enough about these transitions, if we do not cast aside evolutionary thinking like Heidegger did, it becomes reasonable to claim that proposing there to be a single ontological abyss, one that separates a human way of being (i.e., existence) from that of our fellow animals and all other life forms (i.e., captivation) is a misinterpretation of the ontical record. Instead, the evolutionary record documents that there have been numerous profound changes in life history that are relevant to understanding changes in *Daseins*' forms, but also that at least some of these changes can be interpreted within the framework of existence *and* captivation, not existence *or* captivation. I use the phrase "profound changes in life history" with circumspection. Such changes are not profound in a quantitative sense; rather, they are profound in that they are the opening itself of new possibilities for animals. Also, these changes do not simply happen *to* animals; they are also something that animals *do* as new forms of *Dasein* arise. Making such a claim requires interpretation that is partially outside the realm of evolutionary thinking. An additional point is that evolutionists—given their gradualistic presumptions within a strictly ontical frame—do not "see" ontological abysses in these transitions, but neither can they deny them, given the incomplete disclosure characteristic of objective thinking. Indeed, a fuller understanding of the phenomena requires ontological thinking, that is, about the way of being.

So, does evolutionary biology challenge Heidegger's thinking when it argues that the various modern human functionalities—ones that together form a unified being—do not appear all at once? I argue that there is no conflict here. What Heidegger says is that to remove some existentialia from the modern human is to transform that human into something else. But evolutionists address a separate question and do not comment on Heidegger's thought experiment. Evolutionists are interested in the biological origin of the modern human with its (current) suite of characters. These characters are not Heidegger's existentialia, but they (characters) are important in making possible such a further interpretation. That the fossil record demonstrates that characters do not appear simultaneously in (scientific) time does not address who the human being is. We could say that certain fossils are not from "modern humans" or we could give the modern human a separate name, for example, *Homo sapiens sapiens*. But names by themselves do not resolve the issue as to whether fossils are from humans, as Heidegger understands that term. Since Heidegger does not (and cannot) address whether these fossils are from humans, there is no conflict. Even so, evolutionists will, for their own reasons, continue to refer to them as from some sort of human. And people like myself who are interested in way of being *and* in evolution will more easily come to see questions surrounding *Dasein* as more diverse than Heidegger claims them to be.

If we briefly reorient to the other side of Heidegger's abyss, we can ask: How does it stand with captivation? Here, Heidegger is a serious "lumper" in that all of our fellow animals are given the same way of being; whereas, as regards existence, Heidegger is a "splitter" in separating the human way of being from that of all our fellow animals. Of course, he does lump all humans, as least for *Dasein*—versus his understanding of *Da-sein*. Does Heidegger present a philosophical justification for all this, or is he simply expressing in philosophical terms the largely unchallenged and widely occurring anthropocentrism embedded in Christianity and Western society at large? Firstly, I note that Heidegger does not claim to be in the business of justifying, at least when thought in terms of logic or of scientific reasoning, since these apply, according to Heidegger, only to beings. Instead, he is trying to interpret (think) being. Secondly, Heidegger does acknowledge that there is a large diversity of animal behaviors, and we now know that there is a correspondingly large diversity of sensory abilities (e.g., Yong 2022), indicating that different animals interpret different worlds. But here I am using the terms "interpret" and "worlds" as basically naturalistic terms, not as the ontological terms that Heidegger claims are applicable only to humans. So how can we investigate Heidegger's ontological claims if the abyss and the terms underlying it are integral to the claims?

As finite beings, we cannot interpret being as separate from beings. Thus, Heidegger is careful in *FCM* to turn to the phenomenological animal record ("data") for corroboration, while still emphasizing that these data necessarily include an implicit interpretation of being, one that cannot simply be "read" from the data themselves. But Heidegger never challenges the data (e.g., those that von Uexküll cites) as data. Instead, he assumes their correctness and proceeds with his own explicit interpretation of them. But—technical issues aside—could he have done otherwise? Could he have claimed that it is misguided even to think about scientific data? What would such a claim have meant for the possibility of talking about (ontical) animals? And what is the role of this ontical record given that it can be interpreted so differently? Here I would claim that data are particularly useful when they jar our presumptions. If they are not doing this, then it is important that we keep gathering new data, even though data require interpretation (i.e., the hermeneutic circle). If new data *cannot* jar us, then it is possible that we are not interpreting; instead, we are being dogmatic. And how can we know other than by gathering more data? While the question as to how many data are enough is unanswerable, a useful rule-of-thumb might be that the more fundamental the claim, or the more important the claim is for one's *Dasein*, the more we should keep gathering data. But Heidegger devotes only scant attention to the ethological record, and I think this reflects his relative disinterest in animal being itself. Rather, he is more engaged with others who think about animals. That I see as a problem for interpreting an animal's way of being. To remediate this problem (of course, never completely), the remaining two chapters include my (highly selective) interpretation of old and new ethological data that have become available since *FCM*. In doing this, I am imagining a dialogue (of course, through the scientific literature) with our fellow animals, something that Heidegger misguidedly thinks is nonsensical. Particularly for animals living today, I find the possibility of an actual dialogue with our fellow animals to be eminently reasonable. But to claim that a dialogue is even possible indicates that (at least some of) our fellow animals can also think. Here, I am not willing to concede Heidegger's claim that animals cannot think, nor do I concede that I am using the term "thinking" radically different from Heidegger. Support for my own claims first requires my interpreting the ethological record.

I conclude my broad discussion of evolution by returning (see my Chapter 2, Section B: "*Interpretation of Nietzsche's Second Untimely Meditation*") to Heidegger's distinction between "history" (*Historie*) and history (*Geschichte*).[5] Why—you might by now be wondering—do I write so much about ontical evolution when Heidegger's primary interest is ontological? And here is the link in three steps which, for purposes of clarity, emphasizes human history/"history": First, Heidegger thinks that "history" is derivative on history, but this must then mean that "history," while misguided with respect to Heidegger's ontological thinking, nevertheless, can be connected to history. Here, I interpret "misguided" in the following manner: "History," as the naturalistic study of humans through the past events in which they were involved, leaves out the very happening of those events in their meaning to human *Dasein*'s ownness (self) as interpreter. In providing only a partial understanding of history, "history" is not misguided in the sense of being incorrect. But if this is so, then "history" can remain an illuminating, albeit incomplete, science when coupled with *Dasein*'s deeper understanding of history. Heidegger says, "Where man lives without history, historiography [also 'history'] cannot become meaningful for him and hence cannot possibly gain power over him" (Heidegger 1994: 80). In other words, "history" too can become meaningful.

Now to the second step: Human evolution can be connected to "history," and vice versa, without reducing one to the other. This claim about "history" and evolution is based on the now widely acknowledged fact that humans both have a "history" and can evolve, and that while these two are typically talked about differently, what oftentimes (but not always) distinguishes them are distinct temporal frames and methodologies. Any claim of fundamental difference between "history" and evolution should be rejected. Evolution and "history" are always happening and, indeed, they continue to happen today. The study of evolution can be useful in the study of "historic" events (e.g., the migration patterns of ancient humans), just like "history" can illuminate evolution (e.g., consider viral pandemics). Using somewhat different words to say the same, the human gene (biology)/culture ("history") dichotomy is now largely rejected as unwarranted dichotomous thinking. One of the important insights following from this is that human "history" must also have happened hundreds of thousands of years ago. But there is a caveat to this: the study of "history" and evolution must be properly grounded ontologically, and Heidegger is right that scientists are seldom aware of or concerned about this. Nevertheless, I claim that proper grounding can hold if "history" and evolution are properly conceived. Heidegger thinks that history happens in events of (unexplainable) appropriation (*Ereignis*), while "history" and evolution are premised on continuity and the possibility of explanation. But "historians" and evolutionists do not have to deny such events of appropriation (or history, or historical being) in order to talk about continuities that occur outside these events (Polt 2006). With a proper underlying understanding of being, suitably interpreted "historical" continuities too can make important contributions to our understanding of one's way of being. That Heidegger does not extend his acknowledgment from "history" to human evolutionary studies reflects his own unfounded claim that evolution cannot be ontologically grounded in a satisfactory manner. Of course, many findings by human evolutionary biologists (together with "historians") post-date Heidegger, and it is not clear how he might have responded, for example, to the discovery of putative Neanderthal artwork and burial practices.

If the link between "history" and evolution can be made for humans, then why should there be a problem with extending this to our fellow animals, unless our fellow animals can have neither a "history" nor an evolution that relates to their history? One reason that this does appear to be a problem is because scientific studies almost inevitably talk about animals collectively (e.g., as a population), even if—as for many ethological studies—conclusions are based on the behavior of individual animals. This failure to acknowledge the individual animal largely squelches the further idea that an animal can be a self, which is a key condition of there being history in Heidegger's sense. But an absence of acknowledging individuality is not an acknowledgment of absence. Even if evolutionists are not studying these individuals as individuals, nevertheless these individuals still lived their lives—just as now-living animals do; our fellow animals may have history, even if biologists (largely) ignore this.

Now to the third and final step: If "history" and history are connected, and if "history" and evolution are connected, then history and evolution can also be connected. This connection is not based on an underlying necessity (i.e., on a causal inference); rather, it remains one of possibility. I realize that Heidegger would dismiss my syllogistic thinking as misguided logic and inappropriate for ontological thinking. But as regards this particular syllogism, I think Heidegger is on the wrong side of history, and I hope the remainder of this book will prove persuasive. But even assuming that our fellow animals can be in *Dasein*, the task of connecting history and evolution is not straightforward because Heidegger's understanding of history seemingly bears little resemblance to the current study of evolution. Of course, the same is true for his understanding of history and "history." While (ontical) evolutionary records are generally depauperate relative to many "historical" records, they are not—I have argued—fundamentally different. Whether in the recent past or the deep past, our fellow animals were still acting out their individual lives while still open to change, even if we cannot see this directly from the ontical record. Here I invoke Heidegger in *FCM* when he said that the ontical record, while able to provide support, must still be considered secondary in making ontological interpretations. I am making an interpretation, and whether the record is "historical" or evolutionary, it makes little difference. There is no need to invoke an evolution and an "evolution" to parallel history and "history" because "evolution" and "history" are not fundamentally distinct.

Section B. Presentation of My Own Ontological Claims

The study of evolution is the study of transitions that have opened up new ontical possibilities for all life forms. In this chapter, my aim has been to make this understandable and to highlight some transitions that majorly affect the life history of animals within a particular lineage. We might reasonably ask: Does the study of life history, as objectivist (including a third-person viewpoint) and ontical, have anything to do with understanding an animal's way of being? The answer: Not directly. Nevertheless, life history—as deciphered through ethology and evolutionary biology—is relevant to interpreting *Dasein* (including the existentialia) as the ontical being whose very own being matters to himself/herself. However, in saying this, I am claiming that there are animals in addition to humans who are in *Dasein*, and who therefore have existence as their way of being too. Alas, the study of evolutionary ethology and ethological paleontology is in its infancy—indeed, the ethological study of most animals, extinct and living, has hardly begun—and in this chapter I have presented little more than a justificatory framework for thinking ontologically about animal evolution. But in the next two chapters, I turn to ethological studies of individual, living animals to more fully justify my claim that our fellow animals are in *Dasein*. Before doing that, I want to conclude this chapter by presenting my own six theses (see List 3.1) that directly respond to Heidegger's three ontological theses in *FCM*. Like Heidegger's, my own theses are comparative, but thought more deeply as I have tried to do in this chapter, that comparison can be enriched through evolutionary thinking. Like Heidegger, I will not be able to support all of my theses in this one book

....

List 3.1. *Six theses concerning the ways of being of our fellow animals, humans, and other entities*

1. Existence is a way of being of humans and (some of?) our fellow animals. Existence entails that a world opens up for them. Whether existence is also found outside animals—not a question that I engage in this book—is less clear than for our fellow animals, but it is not unanswerable and it is important to address.
2. Captivation is a way of being of all life forms, including our fellow animals and humans. As per Heidegger, creatures in pure captivation are never in a world. Viruses are reasonable candidates for being in pure captivation.[6]
3. Combining #1 and #2, we can say that some animals, including humans, have *two* ways of being—existence and captivation. Individuals of these animals live in and on both sides of Heidegger's ontological abyss.
4. A stone's way of being is not captivation (or existence) because the stone is an assemblage, not an individual, at least in a Darwinian sense. A stone's way of being is relative to one who exists. The same holds for other, more complex nonliving entities, such as rivers (suitably understood), clouds, and stars, but what is "living" and what is "individual" varies across cultures. I do not engage cross-cultural understandings in this book, but neither do I wish to diminish them.
5. As an ontical being with existence as his/her exclusive way of being, *Dasein* comes in various "forms" (e.g., see *Being and Time*, pp. 76–7). Evolutionary and ethological studies can assist in identifying forms of *Dasein* found in our fellow animals through an interpretation of their existentialia. These forms are unlikely to be sorted into rigidly defined taxonomic groupings.
6. *Dasein*'s way of being is never captivation. Therefore, animals—including humans—who live in and through existence are never fully in *Dasein* because they also live in and through captivation. To quantify the different ways of being an animal—for example, as more or less in existence (as per *Dasein*) relative to captivation—is an unwarranted homogenization; it confuses the ontological for the ontical

My theses continue to support Heidegger's claim of an ontological abyss between existence and captivation. However, this abyss no longer separates humans from all of our fellow animals. Instead, humans and our fellow animals are always in both ways of being. As a consequence, we are never fully in existence, or *Dasein*. Instead, we live on both sides of the ontological abyss and that is who we and our fellow animals "are."

Through these theses, I am also proposing that *Dasein* comes in a plurality of forms (e.g., human, bee, octopus—see Chapters 4 and 5). These forms represent different ways of thinking *Dasein*'s unitary way of being as existence in a manner that makes sense of biological (ontical) diversity. The task of articulating individual forms will come through identification of their existentialia (see List 1.1 for human existentialia), and ethological studies have much to contribute to this task. We should be open to the possibility that Heidegger may have underreported some existentialia—particularly ones found in our fellow animals—and warped our understanding of other existentialia so as to exclude our fellow animals from *Dasein*.

Some individual forms of *Dasein* are likely to roughly overlap single higher-level taxonomic groupings (e.g., genus, family, class), thus correlating with genealogical continuity. Others may span groupings that map discontinuously to the evolutionary tree, paralleling an evolutionist's understanding of convergent evolution. Still others may correlate only with an individual species, or even with divisions within a species. An evolutionary inquiry into *Dasein*'s forms will expand Heidegger's *Dasein* thinking by pointing to the connection between *Dasein*'s existentialia and the location of their behavioral correlates on an evolutionary tree. Of course, this still leaves us with a task: We must articulate *Dasein*'s various forms by paying close attention to the diversity of animal life histories and evolutionary changes in them.

Chapter 4

On Bees: Rethinking Heidegger's Claims

Section A. On Bees and Heidegger

In this chapter, I critique Heidegger's own interpretation of ethological studies of bees[1] (Sections A, B). Then, I describe additional ethological studies and I provide my own interpretation within the context of bee *Dasein*, using as my template Heidegger's articulation of human *Dasein*'s way of being as existence through temporality and spatiality (Section C). In the next chapter, I consider a more taxonomically diverse array of animals, but each necessarily in lesser depth than for bees. In both chapters, I aim to support the claims of animal *Dasein* and existence as her way of being. Captivation as a way of being remains important since Heidegger's abyss remains (see List 3.1). But Heidegger does not deeply pursue captivation, and neither do I. It remains one of several important topics I identify in this book that are worthy of additional consideration.

As I turn to the ethological studies of bees, it is important to keep in mind that such studies can only be considered holistic (and therefore make it possible to point to *Dasein*) relative to more reductive fields of biology, for example, biochemistry.[2] Furthermore, ethology remains a separate discipline from (human) anthropology and (human) sociology, and this implicitly continues to reinforce Heidegger's abyssal distinction. A separate problematic feature of most ethological studies is that they typically talk about *the* animal, even though it is the behavior of individuals that is studied. Shared behaviors among multiple individuals are oftentimes aggregated to enable statistical analysis, and non-shared behaviors are ignored as "noise." Fortunately, this is changing (e.g., see "The power of personality" [Pennisi 2016] and "Stochasticity, individuality and behavior" [Honegger and de Bivort 2018]). Just as Heidegger (in *Being and Time*) is not writing about · the human · in general, so I am not writing about the animal in general. Instead of writing about the human, Heidegger points toward what it *means to be* human *Dasein*. Second, his writing points to human *Dasein*, which is not to be confused with the common (scientific) understanding of the human. And third, a proper understanding requires that human *Dasein* be "in each case mine," not human *Dasein* in general. These potential misunderstandings must also be kept in mind as I discuss animal *Dasein*.

In *FCM*, Heidegger devotes more than a dozen pages to the description and interpretation of honey-bee behavior, more than for any animal other than humans. We might wonder why he felt that honey bees, rather than, say, chimps or dogs, were particularly suitable for his ontological analysis of animality beyond his familiarity with Karl von Frisch's influential studies of bee behavior and von Uexküll's popularization of them. One reason Heidegger states is that the study of creatures who are behaviorally remote from the comportment displayed by humans can better highlight the "peculiar character of [nonhuman animal] behavior" (*FCM*: 240–1). Bees do fulfill this ontical criterion. As examples: (1) Bees, unlike humans (but see Wang et al. 2019), are sensitive to magnetic fields. (2) Bees (like many other arthropods) are sensitive both to ultraviolet light and to polarized light, which brings forth floral patterns in nature that remain invisible to human eyesight; however, unlike humans, bees do not see the color red. (3) Like other arthropods, bees have head appendages called "antennae" that are without parallel in the human. These antennae sense electric fields during foraging, they serve as mechanosensors both in the hive and in flowers, and they sense odor and taste. (4) Relative to humans, bees have an unusual eye structure that alters how objects are seen. These, and many others, lead to the conclusion that a bee's life must be very different from that of the human. But, such differences do not by themselves establish that bees cannot also be in *Dasein*. Having distinctive sensorial experiences does make the claim of being in *Dasein* more challenging to establish, but it is not an argument against that claim. How can we know whether bees—that is, bees as individuals—have their own form of *Dasein* unless we seriously engage and interpret ethological studies?

Heidegger's fundamental claim is that our fellow animals are *ontologically* distinct from humans. To make this claim by interpreting (ontical) behavior—as he does in *FCM*—he would have been better served to choose some animal, like the chimp, whose behavior is strikingly similar in some regards to human comportment, and yet to convince his readers that any such ontological claim of similarity is misguided. This is a challenge Heidegger chooses not to make. There is a second reason that Heidegger emphasizes bees beyond their distinctness. Bee behavior, like that of other hymenopterans (bees, wasps, ants, sawflies), is oftentimes described as instinctual. But, is all bee behavior instinctual, and does instinctual (an ontical/scientific term) necessarily indicate captivation? Absent a clear understanding, describing an organism's repertoire of behavioral responses as instinctual should not necessarily indicate a lack of awareness. Instead, we should maintain an openness to there being multiple ontological understandings when biologists use the word "instinct." But what about Heidegger? Might "instinct" have provided him with a politically convenient, naturalistic word to link to his own ontological idea of a whole animal's "darkness" or "absorption in itself"? But any such linkage is misguided—as Heidegger would already know—because any "darkness" conferred by (ontical) instincts must be a different one from the ontological darkness of captivation.

In insisting on a "universal determination of the essence of animality" (*FCM*: 243) rather than, say, species-by-species or individual-by-individual determinations, Heidegger opens himself up to ignoring what bees are doing. In 1929/30, Heidegger's acceptance of the purely instinctual behavior of bees was a mainstream view, and he seems to have simply translated that view into a claim about animal being, one that also conveniently amplified his view about the uniqueness of human *Dasein*'s own being. That humans, too, might have instincts is not an issue that Heidegger engages despite its ontological significance.

One aim of this chapter is to challenge the idea that bee behavior can be understood as "purely instinctual," particularly in the sense that her "decision-making" is bimodal. But bimodality itself addresses a largely methodological point. The deeper question is whether decision-making by bees results from an awareness of their own world and from concern for that world. For Heidegger, bees cannot encounter entities as ready-to-hand or present-at-hand. But how does Heidegger come to this conclusion other than by interpreting the ontical record, and has he done that adequately? Absent Heidegger's own clear guidance, I offer my own best approximation—not an original one—for how to interpret *Dasein*'s ontological awareness in light of the ontical record: if the behavior under study is complex or multifaceted, then there is an increased likelihood that the bee is indeed aware of that very behavior. But even when the behavior is not complex, we cannot be sure that there is no awareness. My claim is probabilistic, not determinative. Neurobiological testing might eventually be able to lead to a determinative claim—as, for example, when scientists conduct brain-scans and conclude that a person is or is not in a vegetative state—but this approach is currently speculative, and I do not pursue it, in part because its reductionist approach presents us with further ontological challenges. A related question is whether the bee is aware of her own awareness, as seems required for there to be bee *Dasein*. In offering a guideline for determining bee awareness, I admit upfront that—like Heidegger—I am privileging ethological research over other ontical approaches, particularly the more reductionist ones, but still (and to repeat), I do not claim that my ontical understanding determines the ontological claim. I further admit that I am making an even deeper claim: in order to talk in an ontologically meaningful manner, there must be some ontical understanding (e.g., a test) that partially informs our ontological interpretation. Science offers one general way of testing, but there are other, mostly less explicit, ways which I have not chosen to explore. For example, later Heidegger turns to the works of "great" poets. But even emphasizing scientific ways of testing, my proposal remains open to bringing in nonscientific, but still ontical, understandings that inform our ontological interpretation, for example, a concern for thriving within local communities or other optimality criteria.

Section B. Reinterpreting Bee Behavior as Presented in Fundamental Concepts of Metaphysics

In *FCM*, Heidegger examines bee behavior in §§46, 59, and 66. In §46, Heidegger says that the bee has (1) its hive with cells, (2) the blossoms it visits, and (3) the other bees of its swarm with which it interacts, but "it does not know the stamens of these blossoms as stamens, it knows nothing about the roots of the plant and it cannot know anything about the number of stamens or leaves, for example" (*FCM*: 193). As for knowing nothing about the roots or about the number of stamens, numerous human children and some adults are likewise unlikely to be botanically knowledgeable enough to pass Heidegger's test here. Heidegger might respond that, in principle, adult humans could pass this test, while it is simply unimaginable that bees could. However, my response is that if knowing the stamen number is important in the life of bees, then they are likely to learn this. Absent such mattering, no conclusion about any supposed inability should follow.

What about Heidegger's claim that a bee cannot know a stamen as a stamen? Again, my response: In order to thrive and survive, a bee must "know" various parts of the flower and be able to generalize to other members of the same flower species, and even to other species of flowers, or else she will not be a successful pollen and nectar gatherer. Bees must "know" that stamens in a particular species of plant are associated with the flower petals in a particular configuration that can vary from one plant to another in order to successfully maneuver through that space. We could say that bees must have functional and topological knowledge of stamens in order to extract food from them. Whether this practical knowledge is also conceptual depends on what one means by conceptual, and in particular its relationship to language. Even Heidegger argues ambiguously on this relationship. Regardless, Heidegger did not prioritize the conceptualization of some life situation over meaningfully directed action, so any strong dichotomy between humans and bees as regards Heidegger's example of the stamen seems unwarranted. Of course, Heidegger would have restricted the word "action" to the human, leaving the bee with mere behavior, but this anthropocentric, etymological maneuver should be rejected.

In §59a, observational and experimental results are described surrounding two clusters of bee foraging-related behaviors—feeding as such and flight orienteering to and from the feeding site and the hive. Then in §59b, Heidegger explains how these observations support his interpretation of the bee's way of being as captivation. I begin with §59b to put Heidegger's thinking upfront. Heidegger argues that in flight, the bee knows nothing of what it is doing as it orients in relation to the sun, as it journeys for a specific period, and as it readjusts its flight pattern in response to visual landmarks and perceived odors. By "knows nothing," Heidegger means that the bee neither recognizes, apprehends, nor reflects on entities in its environment. (Of course, we might, in turn, wonder what these three new verbs mean to Heidegger—whether they have some special meaning.) Rather, Heidegger says, the bee's behavior indicates "a being given over to" various disinhibitors (e.g., sun, time, landmarks, odors) precisely because of the bee's drivenness with respect to these disinhibitors. Indeed, Heidegger's use of the word "behavior" *assumes* such drivenness in order to contrast it with human comportment. In her drivenness, the bee is captivated by her disinhibitor. Every "apprehending of something as something [is] withheld from it." Whether the bee is going to a food source (i.e., "driven by the fundamental drive of foraging") or returning home (i.e., "the instinctual drive to return home"), it is always driven and captivated. Captivation is how the bee "relates" to other entities. Interestingly, Heidegger says that captivation should not be viewed as a "rigid fixation." This is because the bee can simultaneously be presented with multiple disinhibitors, resulting in the possibility of multiple behaviors. However, Heidegger is clear that these possibilities are not manifest to the bee herself; such manifestation remains outside even the possibility of the bee's way of being.

Now to §59a and Heidegger's commentary on feeding behaviors. Heidegger asks why the bee lands on a flower and then, after collecting nectar, departs from the flower. Does the bee recognize the nectar as something present and, subsequently, the absence of nectar as something absent, or does the bee have no awareness of this? Heidegger describes an experimental intervention to address this question. When presented with a super-abundance of honey in a bowl, the bee drinks until satiated and then returns to its hive, but when its abdomen is experimentally excised, the bee continues drinking even as the nectar drips out of the bee's posterior end. Heidegger takes this as evidence that the bee does not know the honey as honey, that the bee is simply taken by its food. The intact bee's satiation is a form of drivenness that inhibits the separately driven activity of drinking. For Heidegger, bee behavior is to be understood via the interactions of various inhibitors (e.g., satiation) and disinhibitors (e.g., food). The bee has no recognition of inhibitors and disinhibitors as such, but together, they lead to a connected series of behaviors, or drivennesses. For example, when the intact bee's drivenness to drink is inhibited, she is then taken over by the drivenness to fly back to the hive. Here, Heidegger's story remains speculative without the experiment involving an abdomenless bee. But as regards this experiment, we should first ask whether an abdomenless bee is still to be considered a bee in Heidegger's holistic sense—for example, do we ever see such bees flying around, living their lives (rather than, say, dying as a result of becoming stingerless while defending one's colony—see Section C4)? And depending on one's answer to the preceding question, we should ask, what sort of interpretation is opened up by the experiment itself? Is the experiment designed to understand the suborganismic *mechanism* of satiation, as a biologist would claim, or an organism's being as captivation, as Heidegger claims? At the very least, Heidegger's interpretation of all this is open to doubt.

Now let's turn to flight navigation. From observation alone, we can say that the (female) worker honey bee collects nectar and pollen (hereafter, referred to as "nectar") from flowers and brings these nutritive goods back from distant places to her hive and her hive-mates. To collect nectar, either she has to discover a suitable patch of flowers herself or else its location has to be imparted to her at the hive by other workers. Either way, there are important questions about how all this occurs—about what is mere instinctual behavior and what must be learned. Heidegger's discussion of this is highly limited, but what does he say? He claims that individual bees only ever collect nectar from the same species of flower, although other individuals of her swarm may collect nectar from other single species. However, he does not note that this fidelity to a particular species of flower lasts only as long as those flowers remain a reliable source of nectar; otherwise, she switches. Would not humans follow a similar strategy?

Heidegger also describes several observations and experiments that relate to a bee's flight-orienteering abilities and concludes that the bee does not know space as space, consistent with her captivation. But subsequent to the *FCM* lectures in 1929/30, interpretations of these observations and experiments have been critiqued and considerably complexified by additional experimentation, so I will delay my discussion of bee orienteering until the next section, where I discuss the new findings.

Lastly, let's turn to §66. The bee, like all animals (sea urchins, great tits, squirrels, woodworms, and woodpeckers are mentioned), is captivated with its own disinhibiting ring. Heidegger says that the encircling rings of the various species should not be viewed as hardened spheres, but instead as rings that intersect with each other. Such intersecting applies both within and across species, and it opens up "an enormous wealth of contents and relations which we can hardly imagine." But more radically, Heidegger says that the encircling ring applies uniformly to all members of its species, unlike human *Dasein*'s way of being, which must be "in each case mine." If Heidegger is here claiming that all nonhuman individuals of the same species must respond identically under uniform conditions, then that is simply wrong, as I document in the next section. All this leads me to ask whether Heidegger—with his deep skepticism that a world can be manifest to our fellow animals—could be open to *any* ethological experiment influencing his thinking on this matter.

Bees, as for other Hymenoptera, are widely acknowledged to display numerous instinctual behaviors. There is no reason to challenge that idea. But, is all the behavior of bees instinctual? Are there aspects of bee behavior that do not fit conveniently into instinct? If we want to consider the bee as a bee, and not as a creature that can display only instinctual behaviors, then we need to learn more about their entire life. Fortunately, Heidegger's near-contemporary—Karl von Frisch—trained numerous other bee ethologists, and they in turn trained others, so bee behavior continues to be extensively studied. Almost 100 years after *FCM* was written, there is much to add to von Frisch's and Heidegger's accounts, so to that we turn.

Section C. Further Ontical Studies on Bees and New Interpretations of Their Behavior

In *Being and Time*, Heidegger says that human *Dasein*'s way of being (existence) is grounded in temporality. Therefore, if I am going to extend *Dasein* to bees, I will have to explain how bee *Dasein* relates to temporality. In this section, I frame my discussion of the bee's ethological record in terms of temporality (and spatiality). Very briefly, Heidegger recognizes three moments of originary temporality in his understanding of human *Dasein*. These moments, which are unitary in their threefoldness, correspond to (1) beenness/"past," (2) presencing/"present," and (3) projection/"future." Each moment has a two-part structure, what Heidegger calls an ecstasis—which identifies the "movement" of *Dasein*'s being—and a horizonal schema—which is "where" this movement is going. Here, I very briefly consider each moment in turn: as for its beenness, *Dasein* ecstatically recollects who it has-been (or, when inauthentic: forgets to recollect who it has been), and it does this in-the-face-of-which it has been thrown (i.e., its horizonal schema). Behavioral correlates of beenness—those that enable an ontological interpretation—include a bee's (1) affectability, (2) attunement, (3) bias, and (4) memory. As for its presencing, in a moment of vision *Dasein* ecstatically brings existence into the situation and discloses what is factically possible (or, when inauthentic: does not dwell but instead finds itself in the general position), and it does this in order to carry out some projection. Behavioral correlates of presencing include a bee's (1) awareness of her environment and of her self, (2) her intelligence (distinguishing possibilities from impossibilities), and (3) her ability to learn (including associative learning, generalized learning [cognition], and social learning). As for the projection (and projecting), *Dasein* ecstatically runs ahead, ultimately unto death (or, when inauthentic: awaits), and it does this for-the-sake-of-itself. Behavioral correlates of projection include a bee's (1) goal directedness, (2) imagination, and (3) anxiety. None of these ontological moments leads in a deterministic manner to a particular ontical outcome. Furthermore, it must be emphasized that individual moments remain within the onefoldness of the threefoldness and, thus, should not be seen as distinct or as not interchangeable; indeed, my interpretation of individual ontical behavioral correlates will sometimes overlap multiple moments, for example, in my understanding of "pain" as beenness, presencing, *and* projection.

The ontological framework that I present for bees should not be confused with the ontical categorization of their behaviors, instinctual or otherwise. While it is important to describe bee behavior that is not instinctual—in particular, because Heidegger suggests that instinctual behavior is best understood ontologically through captivation—this is not sufficient. To stay with behavior and its underlying mechanisms would be to remain in the ontical realm. Rather, if we wish to make a claim that a bee can be in *Dasein*—and thus have existence as her way of being—an ontological interpretation is also necessary, one that goes outside the pursuit of scientists. My goal is to interpret a bee's way of being, and what that way—with its biological variability—may mean for the bee herself. And just as human *Dasein* necessarily occurs within a broader social structure—what Heidegger calls a "people"—so we could say the same about bees in the context of their colony and colony-mates. In striving to cleave from Heidegger as little as possible, I want to emphasize that Heidegger's account of human *Dasein* and of our fellow animals is in no need of naturalizing, since it is already natural, not strictly in the sense of biology but in the sense of not incorporating theological (or supernatural) or any absolutely grounding concepts. Heidegger does talk about nonsensuous phenomena, but in an analogous manner, it is not unusual for scientists to posit hidden states of nature that serve to simplify current understanding and to form the bases for further experimentation.

If one's interest in bee behavior is to identify their underlying biological mechanisms, then the body of the worker bee becomes an appropriate target for experimental manipulation. But if one's interest is the way of being of the bee—as is mine—then manipulations ought to be confined to affordances in her environment, leaving the body intact. This is the demand of organismic holism that is implicit in *FCM*, even if Heidegger does not always follow it.

A recently published book—*The Mind of a Bee*—by Lars Chittka (2022b), who is a highly published bee ethologist and ecologist, has greatly simplified my task through its comprehensive but highly approachable experiment-oriented accounting of the bee-behavior literature, largely restricted to the highly social honey bee and bumble bees. Chittka traces his own scientific lineage through generations of mentor/mentee relationships back to Karl von Frisch, thus testifying to von Frisch's lasting influence. So, in this section (C), I summarize many of the studies that Chittka describes, identifying the authors but not their primary references, which can be found in Chittka's book, as well as derivative summaries found elsewhere (Chittka 2022a,c). Studies not mentioned by Chittka are fully cited.[3]

As I've already indicated, in choosing to follow Heidegger I start by using temporality as the ontological framework for an interpretation of ethological studies. I begin by very briefly clarifying my understanding of the relationship between temporality's three moments (beenness, presencing, projection) and the ethological (ontical) studies on bees.[4] As for beenness, the ethological studies illustrate that a bee comes preloaded with biases based on who she has always been and her previous experiences. These biases fit conveniently and appropriately into Heidegger's understanding of attunement in that they should not be misunderstood as momentary and individual psychological states. But for these biases to be actual attunements, there must also be memory, so another part of what ethological studies provide is the experimental demonstration that bees have memories—that they are not simply responding simpliciter to current circumstances. The demonstrations of bias and memory formation in bees, then, provide my interpretative framework for the temporality of beenness. As for the presencing, ethological studies provide experimental demonstration that bees have a complex awareness of their surrounding environment. The idea that bees are totally taken by their inhibitors and disinhibitors without awareness of their environment—as Heidegger says in *FCM*—becomes increasingly unreasonable as bee ethologists document ever more behaviors which demonstrate intelligence and flexibility. But bees also have a sense of self, if this term is appropriately understood, and ethological studies are helpful here too. Finally, bees are social creatures, and their sociality relates to how they learn (i.e., so-called social learning). As for projection, ethological studies provide experimental evidence that worker bees, like humans, are both future and other oriented; that is, individual workers devote their lives to ensuring that their colony—and they with it—have a future. A worker bee's future-directed projects change over the bee's lifetime—from caring for young, to nest building, to guarding the nest, to foraging—and always these projects are done not for a present reward but in order to make possible an enduring future of the colony. The idea that bees are little robots completely ignores this important interpretation of a bee's way of being. A related and further important contribution that ethological studies provide is to demonstrate that bees do imagine that future. About human *Dasein* Heidegger says (*Being and Time*: 264),

> To Dasein's state of Being belong *projection*—disclosive Being towards its potentiality-for-Being. As something that understands, Dasein *can* understand *itself* in terms of the "world" and Others or in terms of its ownmost potentiality-for-Being.

These three moments of temporality, now applied to the worker bee, must be thought in their unity if we are to understand the bee's way of being. So, it should not be surprising that, in what follows, there is overlap in discussing these three moments.

In *Being and Time*, Heidegger also discusses spatiality but within the context of temporality. Later in his career, he elevates spatiality to co-primordiality with temporality through an understanding of space-time, along with (place-based) historical being. In the last subsection (C4), I comment on spatiality in the worker bee.

C1. Temporality and the "Beenness" of a Bee

Heidegger captures the ecstatic alignment of the beenness of originary temporality in the term *Befindlichkeit*—translated as "disposedness" or "findingness." Disposedness underlies an ontical affectability. For Heidegger, affects indicate that an individual has a past that continues into the present and that extends into the future. Having affects is a step in the direction of there being such a thing as a meaningful life, since affects arise from biases or attunements that point toward what matters to an individual. Affects can also be called "moods" as long as they are not thought of as transient, internal states. Affects relate to biased ways of living, and they are present in groups as well as individuals. Indeed, moods may foster both group identities and intergroup distinctions. Importantly, without affects, there can be no action, although "spontaneous" movements may still occur. While Heidegger talks about attunements with respect to human *Dasein* only, I make the case that a version of attunement should also apply to our fellow animals, including bees.

Affects are possible only because there are memories, so we should briefly explore this relationship. Most straightforwardly, one can have a memory of an object or situation, like a bee does of the hive to which she returns. As this example illustrates, such memories are not drawn at random from one's sensory world; rather, they come loaded with meaning, so even simple memories are underlain by affects. As this illustrates, a memory is never simply storage (Heidegger 1976: 150–1).

The relation between a memory and its object can also be less direct. For example, a bee may remember a close encounter with a spider while foraging, and so she becomes wary thereafter, even when navigating in other locations. So, memory is key to understanding affects—and vice versa—and indeed this may seem obvious since a creature without memory cannot become loaded with biases based on its experiences but can only respond to its present condition in the way that it is. It is only in highly artificial conditions, for example, those of a philosopher or a laboratory scientist, where memories may be discussed in such a way as to cloak their affective dimension. Heidegger has convincingly argued that such a way of talking is derived from a prior understanding of already being in an affective world.

If we want to know what matters to a bee—since it is in this that memories and their affective dimensions are instantiated—then we must carefully observe how she lives and must interpret her behavior in a manner that remains focused on the bee herself. Letting the bee speak for herself means observing her with minimal interference and designing experiments that incorporate motivational features that matter to the bee herself.

Worker bees are highly task oriented—they are attuned in this manner—and once their tasks take them outside the hive (e.g., to bring nectar and pollen to the hive and to explore for new feeding sites), remembering places becomes important to their survival and that of their hive-mates. When leaving the hive for the first time, foraging bees find themselves in novel and sensorily diverse habitats—that is, information-rich environments—and they, like humans, must have the means to remember relevant features without simply storing complete "pictures," which would be both unnecessary for their tasks and (presumably) computationally prohibitive given the small size of a bee's brain. So, an important point is that for a bee to navigate her novel environment, learning must be selective and not simply thought of as storage of and access to huge amounts of information. The "beenness," and, in turn, the instantiated affects of the bee play important roles in guiding this selectivity.

In honey bees, foraging novitiates typically start by exploring and memorizing the terrain (e.g., its visual landmarks, smells, tactile cues) around the hive before starting to gather nectar or pollen. Actual foraging may occur only in subsequent expeditions. During this exploratory phase, the bee must remember her own orientation relative to the hive so that she is able to return from kilometers removed. The point in relating this is to illustrate that worker bees are not simply programmed to collect nectar; they must first learn relevant features of their particular environment.

As for *seeing* her environment, the bee has compound eyes which, while outstanding for detecting movement and good enough for resolving nearby small objects like flowers, are much less effective even at moderate distances. However, a large object, like a tree or a hill, might still be sufficiently resolved at a distance, albeit with much lower resolution than our own camera-like eyes provide us. I mention this issue of vision because it might seem to provide a physical basis for Heidegger's idea that the bee cannot know, for example, a hill as a hill, given that the hill might be little resolved from any number of other large objects. But does not this sort of reasoning privilege our own high-resolution vision? Instead, what the bee "knows" is that—if a source of nectar has been found immediately behind the hill—then she has to divert herself from a straight path and fly around the hill, either clockwise or counterclockwise, to get there. To the bee, the hill matters in that it becomes an obstacle to securing food, while for a human the hill may matter—in another context—in that it provides a means to get a scenic view from on top. Obviously, this is a pragmatic and associational understanding of "hill" for both the bee and the human, one that apparently bypasses Heidegger's "as"-framing of mattering, but Heidegger's own framing may presume high-resolution sight, and with it, broad generalizable thinking. But getting somewhere, whether by the bee or the human, does not require this. (Later, I provide evidence that the bee is capable of generalizable thinking, so "as"-framing may occur—in appropriate situations—for the bee too.) A bee's vision does not allow her simply to look from afar over a field in order to locate a promising patch of flowers. Instead, she uses—as I soon document—a different means to assess distant terrain. But at sufficiently long distances, humans navigate by other means as well (e.g., maps, which can either be internal or external), again due to limitations in visual resolution. Whatever the means of guidance, the bee always has to remember her location relative to the hive. And when her flight path must be circuitous (e.g., because of a hill blocking direct access), she must remember the impediment's location. Then—once she is sufficiently close to the flower patch—her vision takes over from other means of guidance, and with it, her visual memory. In part, bee/human differences result from their highly divergent sensory systems, like their eyes, that lead to different affects (e.g., a bee's compound eyes allow her to be more keenly aware of movement, including the normalizing activities of her hive mates and the threatening presence of predators). To understand the bee on her own terms, then, we need to know how the bee herself senses her environment.

In his description of the bee's waggle dance as a symbolic act which transmits a flower-patch's distant location to members of a hive, von Frisch emphasizes that this is accomplished through stereotypical behavior, which requires of the dancing bee that—in the darkness of her hive—she communicate to her hive-mates an angle (i.e., the angle that subtends the sun's vector and flower-patch's vector from the hive) and a distance (i.e., the distance from the hive to the flower patch), thereby generating a vector from the hive to the flower patch.[5] This vector is interpreted as such by the audience of worker bees through their direct antennal contact with the dancing bee. In addition to the waggle dance routine, there is also the shaking signal—which conveys the need for more foragers to exploit a particularly abundant resource—and the tremble dance—which, alternatively, conveys the relative need for more food processors at the nest (Hölldobler and Wilson 2009). The quality of the food resource is also encoded in the variable number of repetitions of the dance and the variable time between the end of one dance and the onset of the next (Seeley 1996). All this suggests that bee communication reflects sophisticated decision-making that entails both stored representations as well as rule following.

Nevertheless, von Frisch argues that the waggle dance is instinctual, given that it occurs across various species and imposes only modest computational demands, that is, calculating one angle and one distance. But complexifying von Frisch's understanding is the observation that bees of one species can decode the somewhat distinct dance acts of another. Furthermore, the worker bee that originally transmits information can soon thereafter become a receiver of information from another bee whose dance indicates an even more suitable foraging site, illustrating a behavioral flexibility when new information from other bees become available. Is this instinctual as well? And, as more and more simple and not-so-simple bee behaviors get documented, one might start to question whether the assumption that all these behaviors are instinctual is really simpler—and, thus, less demanding of neuronal (computational) capacity—than a model based on generalizable knowledge. Perhaps we need to confront an anthropocentric bias which presumes that "higher" cognition is the sole domain of the human. Furthermore, we need to see that oftentimes seemingly instinctual responses become coupled with more complex behaviors; for example, consider human cultural embellishments that surround our own eating and sexual activities. For the bee, encoding the dance is only one aspect of a successful foraging expedition. Earlier, she is required to remember key features in her complex environment (i.e., topological features that supplement successful navigation, the presence of potential predators such as spiders and birds, the quality and quantity of the food resources that she has discovered, and so on). That all this could be simply instinctual is beyond reasonableness.

And then there are the dance's instructions themselves. These simple instructions do not include the hill that blocks direct access to the flowers, so interpretation of the instructions requires a certain flexibility. Likewise, bees in the hive may also have picked up a particular scent from the dancing bee while the latter was, say, inside a flower, and this scent may provide essential information about the more precise location of the flower patch than what the dance's geometric coordinates provided. So the bee who follows the dance's instructions will still want to be aware of the scents in her environment as it is traversed. And once the bee arrives at the flower patch with its spatially complex distribution, she still has to have some nectar-gathering strategy, while remaining vigilant of predators who lurk near the flowers. Even accessing the nectar within the flower presents challenges. The internal structure of a flower can be like a maze, which requires the bee to learn how to navigate within the flower itself. And of course, flower morphology differs across species, so learning how to do this is part of being a successful forager. That all these could be purely instinctual is likewise beyond reasonableness.

So far, I have described typical worker bee behavior that involves no experimental intervention. Properly conceived, noninvasive experimentation can bring further clarity to understanding a bee's capabilities, motivations, and important features of her environment. So, I now turn to a series of experiments that shed light on our understanding of a bee's beenness, largely within the framework of establishing that a bee has considerable memory capacity. Underlying many of the experiments is the observation that worker bees are highly task oriented, and their task is to figure out how to bring nectar and pollen back to the hive. Biologists have taken advantage of this by connecting the completion of some task—natural or non-natural—that the bees must learn to a sugar reward earned upon completion. For example, Randolf Menzel demonstrated that honey bees learn to associate various colors with sugar rewards, and, when presented together with a blue-violet colored object, bees can establish this associational memory after only one reward, while with a turquoise-colored object, approximately ten rewards are required before the task is learned. Still, bees were the fastest of eleven animal species tested, with human infants being the slowest. Other experiments (in Chittka 2022b) suggest that the bee's differential color receptiveness fits well with what is actually needed to detect and recognize flowers in the wild. An interesting aside is that evolutionary reconstruction shows that it is the plants that initially adapted to the bee's already present color vision, and not the bees whose vision sensitivities adapted to the flowers' colors. We could say that plant evolution initiated the bee's shift from ancestral predation to their present-day herbivory, but that its elaboration over evolutionary time was a co-evolutionary process. So what have we learned? Certainly that flowers and their colors are important to a bee's way of being. But also, we have learned that for this to occur, a bee must be able to remember her past experiences. Of course, making this associational connection between color and reward could be viewed as instinctual, but not necessarily as *purely* instinctual. It is also important to note that what is currently an instinctual behavior could very well have been a learned behavior earlier in evolution, one that became so invariably important that it became genetically fixed. Seen in this light, the distinction between learned and instinct is one of conceptual framing. But it is also clear that even forming associations can indicate learning. For example, Erika Dawson and Ellouise Leadbeater discovered that honey bees learn to associate a colored light (as an arbitrary stimulus) with release of alarm pheromone, which is usually a warning that the hive has been attacked. After sufficient training, the bees respond to the light even in the absence of alarm pheromone; that is, it has become a learned response to an arbitrary, non-natural signal.

Beehives are of particular interest to ethologists because they provide a material record of bee behavior. The hypothesis that the construction process is based on instinctual behavior is reasonable, given the hive-to-hive similarity within a species. But one of the most striking features of the hive—its mostly hexagonal, close packing of the cells—may well be based on the material properties of near-molten beeswax rather than on the bee herself shaping the final cell structure (see Karihaloo et al. 2013). Furthermore, there is considerable evidence that bees must learn—and thus be able to store memories—if the hive is to be properly constructed. For example, bees raised in artificial circular cells (as opposed to the typical hexagonal ones) with no prior exposure to experienced workers still construct hexagonal combs, but the comb cell diameters vary widely and the comb itself is not regular.

There are other less radical (for the bee!) experimental approaches that demonstrate bee learning. François Huber also explored bee flexibility with respect to hive construction. For example, when bees were prevented from attaching the nascent hive to a ceiling and, thereby, from building the hive downward, as is normal, they instead built from the bottom up, thereby reversing some of the normal routines involved in hive construction. Indeed, they even extended the hive laterally. In another experiment, when workers encountered a glass plate (introduced by Huber), which is a poor substrate for hive attachment, they reoriented their building away from the plate, doing this even prior to reaching the glass plate, thereby indicating foresight. Lastly, Huber forced bees to attach their combs to glass, and when one comb eventually became partially detached, the workers not only reinforced the detached part but also reinforced other parts, again indicating foresight. From this discussion of hive construction, we have learned that even the most seemingly stereotypical of behaviors—hive construction—requires the intelligent, learned, and flexible implementation of behavioral routines, and also that some features previously attributed to animal behavior should instead be thought of as higher-order properties of the hive materials themselves.

Thomas Collett and Almut Kelber demonstrated that honey bees remember their environment. To illustrate this they trained bees to visit two identical, but distant, huts, each containing identical interior arrangements of two yellow- and two blue-colored cylinders. In one hut, a reward was placed between the two yellow cylinders, and in the other, between the two blue cylinders. After this training, the bees were again released from their hive to navigate to the huts, but now there were no rewards inside. Once inside, the bees explored each hut differently, depending on its (external) location. Once inside the particular hut, they tended to explore where the reward had previously been placed. The authors conclude that the bees access one memory (i.e., the location of the two huts) in order to access another (i.e., the [previous] location of the food inside the appropriate hut).

We might wonder—as neuroethologists have before us—whether, during their explorations, bees form a cognitive map, which is a mental representation that allows them to "imagine" their environment and to respond in their future actions in ways that instinctual behavior could not. In an experiment that failed to support this idea, Randolf Menzel and his students (plus Chittka) demonstrated that bees did not take a novel (i.e., previously unexplored) shortcut directly between two previously visited feeding sites. Instead, they returned to the hive from one feeding site before going to the other, which was the way the bees initially learned about the locations of the two feeding sites. Of course, taking a shortcut may require motivations other than efficiency (e.g., Might there be increased dangers in the unexplored area? Might there be physical barriers? Might a small error in navigation be problematic? Might the features of the landscape not be conducive for navigation?), and of course a negative result does not establish a lack of capability. Even so, Chittka and Kunze did show that bees are able to take novel return-routes via "dead reckoning"—what Chittka also calls "path integration"—which is the ability to return directly to a site (e.g., its hive) after following a circuitous route. This indicates that bees are able to keep track of—and, of course, remember—their position relative to the hive. However, in this case "remembering" is algorithmically straightforward and would not be compelling evidence for having a cognitive map.

Like humans, bees base their memories on smell as well as sight. To show this, Mandyam Srinivasan and his team first trained honey bees to navigate from their hive to two feeding stations, one of which was rose-scented and the other, lemon-scented. Then, the bees were exposed to one scent or the other in the hive—as opposed to the feeding stations, which were now unscented—and, upon release, the bees navigated to the feeding station that formerly had that same scent, thus indicating that the bees were recalling memories based on scent.

Bees can also remember elapsed time. In one observation, Hugo von Buttel-Reepen discovered that bees remember when one flower species in their immediate environment is producing nectar while another is not. In a second observation, Martin Lindauer found that bees in their hive can dance at night (when there is no foraging) and still accurately indicate the location of a food resource relative to where the sun would be behind the horizon. And in a third, Michael Boisvert and David Sherry showed that bees can be trained to learn actual feeding intervals, varying from six to thirty-six seconds. Most significantly, the bees learned not to initiate feeding behavior—namely, not to stick out their tongues—until the interval was completed. The authors noted that such learned self-restraint has been regarded as indicative of intelligence in vertebrates, including humans. Intelligence is here meant to argue against the idea that feeding behavior in bees is purely instinctual. This experiment also suggests that, beyond remembering elapsed time, the bees are predicting the future.

I have already emphasized that to provide a useful test, an experiment must incorporate motivations for the bee herself, since her aim in carrying out such tasks is not to satisfy the researcher but herself and her colony. For example, bees almost certainly would lack motivation to pass a mirror test,[6] even if they were capable of doing this. We already know a strong motivator for bees, and that is food. Bees are attuned to food, and this attunement is prior to any experimentation. This attunement is not a transient memory but a particular mood. The flexibility that the bee displays in her multifarious ways of obtaining food is not adequately captured by characterizing this as instinctual.

Ellouise Leadbeater demonstrated that the observed behavior of other bees is another motivator for bees. For example, nectar robbing (i.e., accessing nectar from outside a flower by biting a hole at its base, thereby enabling short-tongued bees to harvest nectar from long-tubed flowers, although of course without the pollinating benefit to the flower) can be quickly learned by observer bees. Further, Dave Goulson and colleagues observed that Alpine bumble bees have local traditions of biting holes either in the left side or in the right side of a particular flower species. The interesting point of these observations is that the trait is stably learned despite its arbitrariness (i.e., biting the left or right side), thereby supporting the claim that it exemplifies cultural transmission.

In related experiments, Ellouise Leadbeater also observed that inexperienced bees preferred to land on flowers occupied by an experienced bee, and that these inexperienced bees thereafter remained with the flower species that they had just visited, being tempted to switch only when they observed other bees visiting other species. So, while bees are attuned to other bees—that is, they are social learners—they are particularly attuned to bees who are effective food gatherers. Bees are even attuned to other species of bees, but less so, and much less so for lifeless wooden models of bees that are observed to "perform" the same task. The point is that bees assess the quality of others before trying to copy their behavior.

Bradley Worden and Daniel Papaj have shown that a bee's particular flower-color preference (now, to artificially colored pseudo-flowers) occurs even when the inexperienced observer bees are separated by a glass screen from the experienced bees who are feeding, thus preventing any immediate reward for the observer bee. Upon removal of the screen, the observer bees preferentially go to the same color of flowers that had been previously observed being tasted by the experienced bees. Do the inexperienced bees understand what they are observing, or is this an example of second-order conditioning, that is, associating a feeding bee with a reward and subsequently associating a reward with a particular flower color? This general question is of great interest to many philosophers who study animal cognition (e.g., see the many contributions in Beauchamp and Frey 2011 and in Andrews and Beck 2018 that raise this question), and it is certainly an interesting question from a mechanistic perspective. But from the perspective of the phenomenology of the bee, and her way of being, a strong focus on whether generalized knowledge guides her action or whether, alternatively, it is the making of associational connections, coupled with hedonistic biases, that guide her action, tends to misdirect our thinking toward implicit barrier-setting between us and the bee, thereby shutting down alternative interpretations of bee behavior. The focus of this question—cognitive or not?—is motivated by our own human tendency to prioritize generalized knowledge, and thereby to ignore the importance of making associational connections in guiding human behavior. If we are interested in a bee's way of being, we should focus on the phenomena, and not particular cognitive mechanisms. Furthermore, we should not assume that a bee cannot have a world—in Heidegger's sense—unless it is able to experience within the frame of a generalizable understanding. In the next subsection (C2)—with its focus on the presencing moment of temporality—I present evidence that bees can indeed understand the outcome of a particular task in more general terms, but for now we shall return to reflecting on the moment of beenness through more direct considerations of behavioral biases and memory.

Different biases are apparent in the study of behavioral variability. Long ago, François Huber observed that bees inside an active hive used their wings as fans to ventilate and oxygenate the hive, and that more bees started fluttering their wings as the oxygen concentration decreased. To explain this, Huber proposed that individual bees within the colony are differentially sensitive to the oxygen concentration inside the hive, or perhaps indirectly to noxious smells. In support of this general idea, Karl von Frisch demonstrated that individual bees consistently display different thresholds in their readiness to accept low-concentration sugar solutions and also sugar solutions tainted with unpleasant tastants. Robert Page established that these individual preferences are manifested very early in a bee's life. It is worth noting that different bee castes (e.g., queen, drone, worker) display even more profound behavioral differences, and these differences are the result from different larval feeding regimes by hive-mates, so it should not be assumed that behavioral variability is fully explainable in genetic terms. However, Christian Brandes and Randolf Menzel have generated inbred lines of honey bees and, using various learning tests, have established that learning performance does have a genetic component. Of course, this still does not mean that all variation in learning performance is purely genetic. For example, differential sensitivity to varying concentrations of sugar solutions likely correlates with a large suite of other behaviors, some of which are likely learned. Regardless, what these observations and experiments establish is that individual bees are differently attuned, and that these differences may be important for the life of the hive.

Geraldine Wright tested for different emotional states in bees. In particular, she tested for cognitive bias when the bees confronted an ambiguous situation. For example, honey bees were first trained to associate a 9:1 mixture of two odors with a sweet reward, and a 1:9 mixture of the same odors with a distasteful "reward." Then, she presented them with an intermediate mixture (e.g., 1:1), but first she vibrated half of the bees. The important finding is that the shaken bees were less likely than the unshaken controls to accept the ambiguous stimulus. Vibrating the bees was interpreted as making them more wary of their overall negative situation. In a related experiment, Chittka and his team showed that bumble bees that receive an unexpected food reward prior to engaging with ambiguous stimuli are more receptive than bees that receive no prior reward. Further evidence that bees have emotional responses is supported by the observation that they are attracted to some of the same psychoactive substances as are humans.

Clarification on the usage of "emotion" is needed here. While Heidegger is clear that emotions are not attunements, this is largely driven by his aversion to standard psychological explanations of emotions with their internalist assumptions, rather than to emotions per se, which he certainly does not deny. Thought more broadly, emotions express being-in-the-world, and in this holistic way of thinking, can rightfully be considered as necessarily linked to attunements. In particular, emotional behaviors illustrate that individuals come loaded with biases. It is these biases which show that bees can be differently attuned. So, emotions are meaningful indicators of attunements, although I generally prefer to use the term "affects."

A frequent question about our fellow animals is whether or not they perceive pain. Broadly speaking, I take this to be a question of attunement, since a positive answer indicates that such animals, as who they are, display valenced responses to basic nociception, that is, sensitivity to tissue damage including the threat of it. So what about the bee? Karl von Frisch thought that bees lacked the subjective experience of pain and that they did not even give a reflex-like response to excision of the abdomen. Chittka calls this sort of thinking "a convenient delusion." He points out that while the bee may indeed lack nociceptors in the abdomen, attacks to the abdomen are rare in the wild and, thus, nociceptors in that bodily location may be unnecessary. The situation could parallel that found in the human, who lacks nociceptors inside the abdomen (e.g., consider the ability of large tumors to grow without notice) and in the brain. As an alternative explanation, Chittka says that it is also possible that the huge honey reward that von Frisch provided to the abdomenless bees may have resulted in a "euphoria" that overrode normal responses. Or, Chittka (2022b: 191) says,

> If you are nonetheless confident that a bee does not sense potentially damaging stimuli, try to roll one around between your thumb and index finger, even gently, and see if she doesn't swiftly respond in a manner that indicates that she is not pleased by the sensation (as will become apparent through your own nociception).

Nociception is necessary for survival, but can it happen without pain? Chittka notes that for pain as a subjective experience to be biologically useful, it must be modulated by past experience, attention, and context. A one-size-fits-all response to nociception cannot capture the complexity of an animal's life (e.g., when it is important to ignore the unpleasantness of the pain and proceed, and when it is not). Chittka reasons that if we were to find that bees could modulate their response to harmful stimuli depending on context, then that would provide good evidence that a pain response exists. Indeed, there is such evidence. For example, in response to release of alarm pheromone, other honey bees will protect their hive from bears by attacking the intruder, even though stinging likely results in the death of the bee. The alarm pheromone appears to override the bee's aversiveness to its own bodily harm, perhaps by stimulating the release of painkillers.

There is also evidence for fear-like states in bees.[7] For example, crab spiders hide on flowers and are predators of bees. Through experience, bees learn to minimize their exposure to crab spiders, but avoiding all flowers where spiders might be is not a realistic possibility. Tom Ings and Lars Chittka have explored the impacts of spider attacks on bees and shown that such bees display long-term behavioral changes consistent with what one would expect of an unpleasant, subjective experience. For example, when trained to expect (robotic) spiders on some flowers, bumble bees subsequently spent more time than previously inspecting all flowers prior to landing, thereby reducing their food gathering efficiency. Such bees also display false alarms, where they reject flowers without spiders. This altered behavior lasted longer than twenty-four hours, indicating that the bee had formed a long-term, unpleasant memory.

In my discussion of "beenness," I have emphasized that bees display behavioral flexibility indicative of learning and that this is enabled by storing and recalling memories and manifesting affects (moods), the latter resulting from the bees' various biases (attunements). Thought holistically, memories and affects "influence" how a particular situation is perceived and/or acted on by a particular bee, but even more deeply they enable that situation to be meaningful for the bee.

I want to conclude this discussion of "beenness" with a final example of attunement in bees. I argue (1) that play—as in child's play—is fundamentally an attunement and (2) that bees play. But what is play? To regard play simply as a certain type of mere behavior is to ignore its role in the life of a creature. For example, play has been considered preparatory for learning skills—both social and motor—that may be useful later in life, for example, role playing, fighting, and motor dexterity generally, exploring and inquisitiveness generally. (This latter may be most relevant to bees.) But while this approach to understanding play is informative, it can only be partial because it reduces play to mere utility—usually for the species as a whole—and ends up ignoring the individual being who is meaningfully playing. To remedy this, play should also be understood as a particular way of being for a particular individual that expresses a particular attunement, or predisposition.

To know that a bee is actually playing, we must interpret her behavior. Gordon Burghardt (2005) has usefully provided five criteria of play, but first I note that—typical of how scientists address such issues—"play" is an impersonal behavioral description. Partly this is because scientists tend both to nominalize (i.e., "play" becomes a thing) and to talk collectively about the individual experimental subjects, while I am more interested in "playing" as an active way of being that relates to an individual. Oftentimes, indeed, biologists describe the actions of individuals, but then they go on to treat the results of many individuals collectively so that statistical methods can be applied. Again, this is useful, but it is not the same as addressing what a particular behavior means for the bee herself, even though the general criteria of play must also apply.

If we keep in mind that "play" is both an ontical and an ontological term (e.g., it refers to behavior and to an aspect of a way of being), we can avoid confusion. Here I list Burghardt's five criteria of playful activities: (1) do not directly contribute to survival, (2) are intrinsically rewarding, (3) are distinct in form from other more functional activities, (4) can be repeated but not stereotyped, and (5) occur under stress-free conditions. These criteria are designed to eliminate alternative explanations, and not to define what play is.

With these criteria in mind and under highly controlled laboratory conditions, Galpayage Dona et al. (2022) asked whether bumble bees play. In their first experiment, the bees' nest box was connected via acrylic tunnels to a combined "flight and ad-libitum-feeding" chamber, but to get there, the bees first had to walk through another chamber that contained bee-sized balls. These wooden balls—yellow ones (six total), purple ones (six total), and natural-colored ones (six total)—did not block direct passage to the flight-and-feeding chamber but rather were pseudo-randomly distributed on both sides (three of each color) of the central passageway. All of the balls on one side—and one side only—were immobilized by gluing them to the floor ("the immobile side"). Then, the behavior of individual bees entering the experimental arena was observed. Given that the balls did not block direct access to feeding and flying, there was no necessity for the bees to be diverted by and to interact with the balls. Furthermore, the bees were already well fed and had had previous access to flying, so they were not pre-stressed. The observations: Over a three-hour test period, forty-five bees undertook ball-rolling actions from one to forty-four times. This was repeated during each of eighteen consecutive days, yielding a total of 910 ball-rolling actions. Prior to the first ball rolling, bees showed no preference for going to either the immobile side or the mobile side. However, after the first ball rolling experience, there was a significant preference for that bee to return to the mobile side, suggesting that ball rolling was an inherently rewarding activity. Of note: The bees' ball-rolling behavior does not mimic some adaptive activity—like proboscis extension or biting when feeding—nor do the bees become more efficient in ball rolling, which might suggest a goal-directed activity.

In the second experiment and using the same experimental design, it was demonstrated that younger bees showed more ball-rolling activity, similar to what has been found with other animals that play.

The third experiment was designed to test whether the ball-rolling activity can act as an unconditioned stimulus, as one would expect if ball rolling was inherently rewarding. During the training phase, the flight-and-feeding chamber was connected to the nest box via a yellow-colored chamber with balls and, at other times but for the same set of bees, via a blue-colored chamber without balls. For a distinct set of bees, the color of the chambers with and without balls was reversed. The all-wooden-colored balls were pseudo-randomly distributed throughout the chamber, and the chamber entry and exit were baffled to prevent the bees from seeing the balls until they were inside the chamber. The bees were thus trained to associate the blue-colored chamber with balls and the yellow-colored chamber without balls, or vice versa. Then, during the testing phase, the nest box was connected via an acrylic tunnel to a T-junction that provided access to both the yellow-colored chamber—now without balls—and the blue-colored chamber—also without balls—for both sets of bees. In both iterations, entry to the chambers was baffled, so from the outside the bees could not see that balls were missing from the chambers until they had entered. Tests were performed on one bee at a time. In these testing set-ups, there was no flight-and-feeding chamber. The observations: When presented with a choice of (empty) colored chambers, the bees significantly preferred to enter the one which they had learned during the training phase to contain the balls, although the bias was much stronger when the balls were inside the yellow-colored chamber. Even if we presume that the bees might have been expecting to eventually enter the flight-and-feeding chamber, which was no longer attached, the key observation is that they still chose to go into the chamber that had had balls over the chamber that past experience had shown not to have balls. The conclusion: These experiments argue that the bees find ball rolling inherently rewarding. The authors cautiously suggest that bumble bees manifest a behavior which "resembles object play," which adds to the growing evidence for "a form of sentience in these insects." While Heidegger would be uncomfortable with the word "sentience," given its implication of an isolated interiority, he should no longer be uncomfortable with the suggestion that play is one of the bumble bee's attunements.

C2. Temporality and Presencing of a Bee

Heidegger captures the ecstatic alignment of the "present" of originary temporality in the term *Anwesen*—translated as "presencing" (Capobianco 2021). It refers to the dynamic character of being itself, but also to *Dasein*'s engagement with being as the manifestation and emergence of all entities. The latter, in its nominal form, *die Anwesenheit*, points—for some—to the constancy of what-is-present, which Heidegger calls "beingness," a principal focus of metaphysics. Heidegger contrasts beingness with "letting presence" (*das Anwesenlassen*) and with the failure of metaphysics to acknowledge the "earth" and the "nothing" in their essence, both of which are later attempts to think being separately from beings. Since the current chapter focuses on the way of being of a bee—that is, of an entity—it would be difficult not to claim that my interest is not with the being of beings, but this should not be equated with the constancy inherent to beingness as traditionally understood. The early Heidegger too was largely focused on the being of entities, particularly human *Dasein*.

In Subsection C1, my aim was to use ethological studies to support the idea that beenness is an appropriate way to describe bee *Dasein*. In this subsection (C2), I aim to show that bees can also presence, thereby extending my case for bee *Dasein*. To do this, ethological studies are garnered to support the claim that bees are cognitively aware of their surroundings and of themselves, that they learn through experience, and that their behaviors are too complex to be explained simply by instinctual responses. But any such biological facts still require interpretation, as Heidegger did in *Being and Time*. Human *Dasein* is to be understood not as a set of abilities but as a way of being to which the abilities are subservient. So it must also be for bee *Dasein*.

In Subsection C1, I also suggested that even when a bee's behavior is based on associational memories coupled with biases, the claim that such behavior is instinctual is unwarranted. This is important because instinct becomes, for Heidegger, illustrative of animal captivation, with its various inhibitors and disinhibitors, and little more. Where animals do make decisions by remembering past associations, such memories can be complex, and even more so as the number of associations to be remembered increases. Chittka has argued that as the variety of associational decisions in a bee's life becomes large, more generalized learning can actually be viewed as being computationally simpler than learning by association. Even so, there is ethological evidence that bees do have generalized learning, for example, in terms of understanding the outcome of a particular task. So, before documenting the various phenomenological awarenesses that a bee displays, I address this cognitive issue.

Olli Loukola and Cwyn Solvi trained bumble bees in the non-natural task of moving a ball to the very center of an arena in order to get a reward. Of the three balls in the arena, the two closer to the center were glued to the arena's surface. So, in order to get the reward the bees had to move the farthest one. This they learned, and they learned it faster after observing other bees who successfully completed the task than when they observed the ball moving to the center by itself.[8] (Magnets under the arena were used when moving the ball "by itself.") So, this experiment again reinforces the point that bees, properly motivated, can learn even non-natural tasks.

But do the bees understand what they are doing in this experiment? To test this, the experimenters again introduced observer bees to watch other bees moving the ball to the center. Then, the observer bees were given their chance at the task, but now none of the balls were glued to the arena's surface. What they did was indeed interesting: instead of mimicking the other bees by moving the farthest ball, the bees moved the closest ball to get a reward, and they did this without any prior experience; that is, they did this the very first time they tried. So, they were not simply copying the behavior of others but must have understood the goal of the task (i.e., to get the ball to the center) in a way that was partially separable from the previously observed means for achieving that goal (i.e., moving only the most distant ball). The ability of bees to solve never-before-experienced object manipulations, like learning to roll balls in order to get a reward, should be considered an indicator of generalizable learning.

Of course, any particular behavior can be specified algorithmically (i.e., in a manner, like pure instinct, that invokes no generalized learning or even making associations), but this caveat applies not only to bees but also to humans. Furthermore, the point of having bees perform non-natural tasks is that, while a human can describe the task algorithmically, it is unlikely that a bee instantiates such an unusual algorithmic routine. At the least, the bee would have to form a cognitive connection between the new task and a serendipitously useful instinctual routine. But if the available routines are not useful, and still the bee solves a task, then it becomes hard not to invoke either generalized learning or forming associations.

Part of the argument for generalized learning rests on a bee's breadth of actions that make instinctual behavior and association-driven behavior increasingly demanding of "storage space" in an animal's brain, particularly when that brain is small; for example, the bee's brain is approximately 1–2 cubic millimeters. But another part of the argument against instinctual behavior is that some of the tasks I have described require social learning as opposed to autonomous behavior. Further arguing against instinctual behavior is that social learning among bees is most effective when they observe their own species, less so when they observe other species of bees, and least of all when they observe nonliving "actors."

The famous waggle dance (described in Chittka 2022b)—previously thought by discoverer von Frisch to be a strictly instinctual behavior—has now been shown to require social learning in honey bees if it is to be properly performed (Dong et al. 2023). To show this, Dong observed the dance of bees who (1) discover a distant flower patch but who (2) have never observed other bees performing the dance. Under these conditions, the naive bee's dance imparts highly variable vector information that consistently overestimates the actual distance, in contrast to bees who have previously observed the dance of seasoned foragers. Thus, social learning improves waggle dancing, while also making clear that there remains an instinctual aspect. Having established social learning in bees, biologists might then proceed to offer hypotheses for further testing as to why bees display social learning, rather than purely instinctual behavior. For example, two general possibilities limiting instinctual behavior could be the presence of developmental constraints within a highly variable environment or evolutionary constraints on the rate of adaptation. Such explanations contrast with the study of social learning in humans, which, until recently, has largely been under the purview of disciplines outside biology, namely, anthropology, sociology, politics, and education. In these disciplines, talk of developmental and evolutionary constraints as understood by biologists is typically avoided—sometimes actively so—in order to avoid (at least in part) the social stigma of fostering deterministic thinking, including particularly instinctual behavior. But by avoiding biological thinking, the positing of a caesura between humans and our fellow animals can appear unduly reasonable, and in a manner that evokes and supports Heidegger's own abyssal ontological thinking. (See related, but more expansive, commentary by Bridges and Chittka 2023.)

While the waggle-dance experiment of Dong et al. (2023) has the attractive feature that it documents a normal behavior in a normal environment, the bees in that study are not fully normal in that they have not been exposed to experienced foragers. Furthermore, the outdoor environment, while natural, is complex and possibly variable in ways that might influence the outcome. This is not a criticism; rather, it is meant to point out that conducting a single experiment that eliminates all alternative hypotheses is not possible. With that in mind, I cite another study (Bridges et al. 2023) to illustrate social learning, this time in a subterranean species of bumble bee (*Bombus terrestris*) whose colonies are of modest size (≤ 160 individuals). Here, the experimental task is highly unnatural—learning to rotate, either clockwise or counter-clockwise, the lid on an underlying container in order to gain access to the sucrose solution within—and the environment is highly constrained, that is, within a room. The key experiment is to expose naive bees to a demonstrator bee, one who has been trained to rotate the lid but who does this in only one direction, and then to observe whether the naive bees can likewise learn to open the lid (they can!) and whether they rotate the lid in the same direction as the demonstrator (they do!). As a control, the authors find equivalent results when the demonstrator is trained to turn the lid in the other direction. And as a negative control, it is observed that a few naive bees perform this task when not exposed to a demonstrator. Those that do succeed do so less proficiently, and subsequent learning by other observer bees in the colony is more unstable.

One more example of learning non-natural object manipulation: Sylvain Alem, Cwyn Solvi, and Lars Chittka successfully trained bumble bees to pull a string that extended underneath a slightly elevated clear plexiglass plate. Underneath the plate, the string was attached to an artificial flower with a food reward attached, and to get the reward, the bees had to pull the flower and food out from beneath the plate. For most bees, successful learning either occurred in steps or by observing others, but there were two bees who learned spontaneously. Speaking more generally, Chittka (2022c) says that in performing experiments on learning behavior, oftentimes there are a small number of individuals who solve the tasks more quickly or more efficiently or in unexpected ways, and that these exceptionally innovative bees also display more variable and more exploratory behavior. Chittka links this variable behavior of individual bees with their differing intelligences, and draws a connection with observed variation in the general intelligence of humans. Considered together, these experiments at least make plausible that bees can have generalized learning. As Chittka says, analogous results with nonhuman primates have been understood as illustrating generalized learning, and it would be unfair to the bees to interpret them differently.

A skeptical biologist might wonder whether a bee understands the role of other bees in accomplishing a cooperative task, or whether task accomplishment is a by-product of individual action. This was the question that Loukola et al. (2024) addressed experimentally. To do this, they first trained bumble bees to push a block or a door in an arena in order to obtain a food reward. Training occurred with individual bees and, separately, with paired bees. A key observation for the paired bees was that, after training and when entry of the partner bee was delayed, the pushing task was accomplished more slowly than by the solitary (control) bee, indicating that one partner bee ("the waiting partner") was waiting for the other ("the delayed partner") to arrive before engaging the pushing task. Further supporting this interpretation that their coordinated behavior was socially influenced is the observation that the waiting partner turned around—and presumably looked behind herself where the delayed partner would be expected to be—more often than by chance or than by the control bee.

In the previous subsection (C1), I described experiments by Geraldine Wright and by Chittka and his team which demonstrate that bees can be differently attuned, and that this can result in their making different choices when presented with the same ambiguous situation. In an extension of these studies, Cwyn Solvi and Andrew Barron also used ambiguity experiments, but now, instead of there being only one choice for the bees to make, they were offered (via a reward incentive) a third possibility, namely, to opt out of making any decision. Interestingly, this opt-out possibility was chosen as the tasks became more difficult, "as if they were aware of their own uncertainty." If this interpretation is correct—and ethologists do assume this is the case when apes and dolphins perform similarly in similarly designed experiments—then it would indicate that bees can know what they know, i.e., they have metacognition.

I now describe experiments that illustrate the variety of ways in which bees navigate day-to-day in their complex environments, showing cognitive awareness, flexibility, and the ability to interpret. In one experiment, Lars Chittka and James Thomson trained laboratory-reared bumble bees—that is, bees with no previous experience in the wild or with flowers—to navigate plastic, flower-size T-mazes. The entryway to the T-mazes was located at the bottom of the T and was color-coded with blue to mean "turn right" or yellow, to mean "turn left." This turn refers to the decision the bee must make once she has traveled up the base of the T to where it intersects the top of the T. Only if she makes the correct decision does she receive a food reward. Learning the task takes time and multiple visits. Analogously, it takes an inexperienced forager bee five to ten times longer than an experienced one to navigate the inner spaces of a natural flower. Not surprisingly (or is it?), bees learn to deal with novel flower types faster when the morphology of the new flower is more similar to the old type, implying that previous knowledge is being accessed to solve a new problem. But experienced bees who forage multiple flower types are slightly slower and make slightly more errors than those who remain with one, suggesting that there is still a cognitive burden in learning these tasks.

From all this, we can conclude that bees can learn to handle multiple flower types, but furthermore, that knowledge of these skills is partially retained throughout the bee's lifetime (which is approximately three weeks in the case of workers).

Johannes Spaethe asked how bees focus on a single flower type and—in the moment—on a particular flower within that type, while at the same time ignoring the multiplicity of other flowers and flower types, plus other objects. One partial answer that has already been highlighted is that foraging bees pay attention to other, experienced bees. But do bees really pay "attention" in the sense that they have now learned what they are looking for? To understand this, it is important to recall that bees have two visual systems, one that is in color but low resolution, computationally slow, and useful only at very short distances (depending on the size of the flower), while the other is in monochrome and higher resolution, computationally faster, and therefore useful while flying rapidly at farther distances, but typically not beyond 10–20 centimeters (again, depending on the size of the flower). Spaethe showed that within the shallow horizon of a bee's vision, it was the brain-processing speed—and not potential flight speed—that constrained the bee's search for flowers. What are the behavioral consequences of all this?

First, we could ask how bees use their two visual systems while foraging. Using plastic flowers (with food rewards inside) of varying sizes, Spaethe showed that when the flowers were large, bumble bees used their color system, but as the flowers got smaller, the bees suppressed their use of the color system and switched to using the higher-resolution monochrome system, indicating that the bees have "an attentional gear box" at their control.

Second, we could ask how honey bees gather information while foraging. A honey bee's inability to search the deeper horizon, as humans do with our camera-like eyes, constrains the size of her visual field, and when looking for harder-to-resolve small flowers, her field size must become even smaller; that is, she must fly even closer to identify them. The bee's solution to this constraint is to search her environment in a strictly serial manner—through a series of attentive looks. Again, this can be contrasted with humans, who detect stimuli in parallel, for example, by identifying with one glance the movement of an object in a broad field of unmoving objects. We might say that the end result is similar for the honey bee—she is able to explore a broad field of flowers—but her serial approach takes longer. Perhaps the biological constraint of having a small brain with limited processing power underlies the serial nature of a bee's attention during foraging. However, each of the honey bee's ~850,000 neurons (versus ~86 billion in a human brain) can form up to 10,000 synapses with other neurons—perhaps a billion total—to provide many possibilities for altering information flow. Regardless, these experiments emphasize that when assessing a bee's intelligence, the parameter of time required to complete a task should be incorporated into the study, and some human-centric standard of completion time should not be assumed. The experiments also raise another interesting question: What does a bee see when it is not moving, that is, when it is not looking in a serial fashion? And might there be a necessary linkage between movement and vision (action and perception?) for honey bees, something that is alien to humans?

Facial recognition is critical to the functioning of human society, so much so that a region of the brain (the fusiform face area of the fusiform gyrus) is devoted to this task, and failure to remember faces is considered a pathology (called "prosopagnosia"). So what about bees? First, as an interesting aside about another hymenopteran: *Polistes* wasps—despite having very differently organized and much smaller brain size (relative to humans)—recognize and remember the distinct facial markings of individuals in their own colony, as part of the means for establishing and maintaining intra-colony hierarchies. But, honey bees do not have distinct facial markings, so it is unlikely that they have a region of the brain devoted to this specific visual task. Nevertheless, honey bees can be trained to recognize an individual human face, when—in the training phase only—that face is associated with a sugar reward amidst other faces that are associated with quinine, which is undesirable. Heidegger would rightfully ask whether the bees are seeing the human face as a human face, and indeed that may be unlikely. Perhaps, instead, they are seeing the human face *as* a peculiar sort of flower, illustrating that the bee may still see the human face as something and that unless we acknowledge what matters to the bee any interpretation of her way of being is likely to be misguided.

Another example of complex learning: Cold-tolerant bumble bees must devote considerable energy to maintaining their body temperature, but internal heat does not come entirely from the digestion of their food. Additionally, they prefer to forage for flowers of specific colors whose nectar is relatively warmer than that found in differently colored flowers, thereby garnering a source of "free" heat. This not-so-simple act, with its multiple interactions and rewards (nectar and warmth), can be deconstructed as follows: (1) the bee senses the temperature of the nectar, (2) the bee senses and then associates the color of the flower with the temperature of its nectar, and (3) the bee restricts its later foraging to flowers of the desired color.

Martin Giurfa and Randolf Menzel have shown that honey bees are able to form and follow rules based on the recognition of common features of distinct objects. For example, bees can be trained to distinguish the bilaterally symmetrical objects within an array of asymmetrical ones. Once trained (by linking symmetrical objects to food rewards and asymmetrical ones to the absence of a food reward), the bees are now able to choose symmetrical objects of shapes not seen previously. (The inverse experiment—training bees to select asymmetrical objects—yields comparable results.)

Giurfa extended these studies to show that bees could be trained, again using food rewards, to choose a visual pattern shown previously over a novel pattern, and that this could be extended to new sets of (recognizable and novel) patterns, thus suggesting that bees do not simply memorize each and every individual rewarding pattern but, instead, must have a more general understanding of "sameness/difference" (as for symmetrical/asymmetrical in the previous experiment) that they store in memory. Such an ability in bees approximates a central aspect of concept formation in humans. Equally impressive, bees are able to be trained in this type of "same/difference" experiment using one sensory modality (e.g., olfaction) and then, without further training, to select the same (or different) stimulus using another modality (e.g., sight).

The finding that bees can switch between modalities raises the question as to how one sense (e.g., sight) can communicate with another (e.g., olfaction). Humans, for example, are able to feel an object (its shape, hardness, overall size, etc.) in the dark and then identify that object by sight, and vice versa. One proposal is that humans form a mental representation that is inter-modally operable, but how that might work at the level of the brain is not clear. Experiments by Lars Chittka and colleagues indicate that bumble bees, too, can recognize an object with one sense and then, later, identify the same object using another sense. Might they, too, form a mental representation, or is there a different explanation?

Another cognitive behavior of bees that might have surprised Heidegger is the ability of honey bees to count. To demonstrate this, Chittka and colleagues placed four 11-foot-high tetrahedral tents as markers across a transect in an open field (tents were located at 75, 150, 225, and 300 meters from the site of the bee's release). A feeding station was then placed midway between the third and fourth tents, and the bees were trained to go to the feeding station. Then, without moving the original markers or the feeding station, new markers were added prior to the third marker, and the bees were once again released. Now, instead of flying the same distance to the feeding station, they tended to stop and search immediately after the new third marker, indicating that they were counting the number of markers rather than measuring a distance. Removing one of the original markers (and the feeding station) had the opposite effect—the bees traveled farther to the new third marker.

Howard et al. (2018) tested whether honey bees understand the concept of zero numerosity. First, individual bees were trained to recognize "greater than" and "lesser than," using white blocks that had different numbers of black shape elements on them. (The shapes varied so that the bees were trained to the number of elements rather than to their shape.) When presented with two blocks having different numbers of elements, bees were rewarded whenever they landed on the block with the greater number (for the "greater than" group) or to the block with the lesser number (for the "lesser than" group), and they learned to do this for displays with novel combinations of elements. Next, the experimenters occasionally used blocks with no shape elements on them. How would the bees rank this zero set relative to non-zero sets? Spontaneously, the "lesser than" group of bees landed on this empty block when presented together with a block displaying one or more elements, although the accuracy was greater as the magnitude of the difference increased. Such a numerical-distance effect with the empty block indicates that "zero" is a quantitative concept for the bees. In another experiment, the bees were trained always to select the two-element block as lesser than any other one presented (e.g., with three, four, five elements). Following this, they were given conflicting information: to choose between the two-element block and a newly introduced zero-element block. In that situation, they showed no preference, indicating that the two were both plausible choices. The authors conclude that these experiments show both associative and conceptual learning.

Howard et al. (2019) extended these studies to test whether bees could solve unfamiliar problems that involved adding or subtracting one element from a group of elements. Here is how the experiment worked: In the training phase, bees were shown an initial display with a specific number and color of shape-elements (either 1, 2, or 4 blue elements for the addition tests; either 2, 4 and 5 yellow elements for the subtraction tests), after which time they entered a "decision chamber" in which they were to select one of two new displays, one display with either one more shape-element than the initial display (either 2, 3, or 5 yellow elements for the addition tests = the correct number for this test) or with one less shape-element (either 1, 3, or 4 blue elements for the subtraction tests = the correct number for this test) and the other display with a different number of shape-elements (any number greater than or less than the correct number). The display was "selected" by flying to and landing on it, at which time the bees either received access to a rewarding sugar solution for making the correct decision or they received access to an aversive quinine solution for making an incorrect decision. (Having an appetitive-aversive strategy improved the bees' performance over simply rewarding for a correct answer.) In the testing phase (which utilized bees that had completed the training phase), the tests were without reward or punishment, the shapes of the shape-elements displayed were novel, and the number of shape-elements was fixed at three (because that number had not been used during the training phase). For both the addition and subtraction tests, the bees chose the correct number significantly more often than random. The authors make two general points about this study: (1) Despite their small brain size, bees have the cognitive ability to solve addition and subtraction problems—to interpolate learned operations into partially novel situations. Doing this requires that they be able to represent numerical attributes, and then manipulate those representations in working memory. (2) Worker bees, whose average lifetime is only a few weeks, are nevertheless able to learn numerical cognitive tasks within that timeframe. This ability may be useful to bees in the natural world for assessing and remembering the richness of an area for foraging opportunities. Relatedly, this ability in bees demonstrates that human-style language is not required for such cognitive tasks.

Having described a variety of behaviors which support the idea that bees are intelligent creatures, we can now confront the issue as to whether they have a sense of self. Without such a sense, Heidegger would claim that any behavioral movements that bees undertake cannot rightfully be redescribed in terms of actions (in his meaning of that word). But how might ethologists test for that? One widely cited metric of self-awareness is the ability to pass the mirror test, a test that likely provides no incentive for a bee. Fortunately, a more bee-appropriate test has been devised by Sridhar Ravi and his team. They asked whether bumble bees are aware of their own size as they fly through gaps in a partition. What they observed was that as the width of the gap became less and then approached the width of the bee's wingspan, the flying bee shifted her orientation from flying straight through the gap to angling her body or even flying sideways, thereby reducing her effective width. They also observed that this shift in orientation occurred sooner for larger bumble bees. (Bumble bees from the same colony vary widely in size because, as larvae, they are fed differing amounts.) Finally, they observed that just prior to traversing the gap, the bees momentarily delayed their flight trajectory in order to visually examine the gap. All these observations indicate that bumble bees are aware of their own body's dimensions, which is one aspect of self-awareness.[9] But here a valid criticism can be raised: The implicit understanding of "self" as described in this and the preceding paragraph is both firmly embodied and in the present, while Heidegger's understanding of self in *Being and Time* is that of a future-oriented human *Dasein*, that is, an individual being whose way of being occurs with resolute projection such that its own being matters to himself/herself. To Heidegger, a demonstration that one is aware of one's own bodily dimensions is, at best, derivative of the deeper issue as to whether this "body" can—in and of itself—have a "for-sake-of-which" (*Worum-Willen*) and a sense of meaningfulness, which includes a self-interpretation. To consider this and other matters, we now turn to temporality's third moment—the "future."

C3. Temporality and the Projection of a Bee

Heidegger captures the ecstatic alignment of the "future" of originary temporality in the term *Entwurf*—translated as "projection." For human *Dasein*, the question is whether one "runs ahead, most centrally into death" (as authentic *Dasein*) or whether one "awaits what will happen in our everyday tasks" (as inauthentic *Dasein*), but Heidegger is clear that both entail projecting. The critical distinction is whether a particular projecting onto possibilities is "one's own" or whether it is uncritically adopted, for example, from one's peers, culture, or tradition. "Running ahead" is an ontological proposition; it refers to a way of being—a "movement" that is intrinsic to *Dasein*—and not primordially to a specific ontical project of *Dasein*, although it has also been used in that manner. Projecting opens up what Heidegger calls the play-space (*Spiel-Raum*) for the existential possibility of being there (*Dasein*); that is, in opening up this space, *Dasein* is always ahead of its own actuality, with possibility itself being manifest through "projects" that constitute its very way of being who it is. These projects are not to be understood primordially as what one has previously done (e.g., I am a doctor because I have a doctor's degree) or as what one is now doing (e.g., I am a doctor because I am now treating patients), but as one's way of being that requires an ongoing engagement with the future (e.g., I am a doctor because I have committed—and must continue to commit—my life to helping sick people). Projecting entails being-forward-of-oneself. Even so, while existential possibility has priority over actuality, projecting cannot be manifest without the having-been (or beenness; see Subsection C1) and the "present" (or presencing; see Subsection C2), which all together give rise in their unity to world formation, and to understanding and interpreting entities in that world. But the central issue that still needs to be addressed—and to which we now turn—is whether any of this relates to bees. Heidegger thinks not, but, as I have argued, his anthropocentric approach to explicating *Dasein*, and particularly his fore-understanding of basic concepts (or "formal indications") such as "*Dasein*" itself foreclosed this very possibility.

As we know, the ontological cannot be addressed simply by observation, as though it were an entity. So, we once again turn to interpreting the behavior of the bee. In particular, we ask whether bee *Dasein* is future oriented, as indeed it must be if we are to consider it *Dasein* (see Balter 2013 for a review of numerous animals). So what do we know? One ethological observation that receives overwhelming support is that bees are highly task oriented. Is a bee entirely in the present while performing some task? First, let's consider what it would mean for a human to be in the midst of performing some task. We would not say that she is entirely in the present. This is because a task has a (distant) end, and that end is likely to be part of a larger goal with its own, even more distant end, and so on. Now, in the case of the worker bee, the larger goal of its nectar-gathering task is to sustain the ongoing life of the colony and her members. Such a simple event as a worker bee collecting nectar should not be viewed only as an egoist undertaking, for example, simply to satisfy the bee's immediate hunger, or, as I have already shown, as simply instinctual behavior. Instead, a worker bee's food-searching behavior should also be interpreted as reflecting her own orientation—her own commitment—toward her own future and that of her colony. But what about those highly exploratory worker bees described by Chittka who never collect nectar or pollen? What they are doing is searching for new feeding areas, which, while perhaps not needed in the present and might never be needed, could become important at a later time, so this activity too should be seen as future oriented and entailing an ongoing commitment to the colony. Here, Heidegger would question whether bees, or any of our fellow animals, are aware—or can even be aware—of future-oriented aims. But Heidegger's understanding of "awareness" is not to be equated with the ordinary modern usage of "consciousness" (see Seth 2005), a term he largely rejects as misguided. Instead, in *Being and Time* Heidegger describes a carpenter who is aware of the hammer but who is not necessarily conscious that he/she is using it. (Heidegger does not evoke consciousness; that is my scientific translation.) So, for Heidegger "awareness" can be practical. Once "awareness" is no longer necessarily, or even primordially, to be thought of as a particular sort of mental act, it becomes harder to claim that our fellow animals cannot be aware.

Heidegger claims that bees live in a kind of darkness which extends from their past through their future. In the preceding two subsections, I have tried to undercut this idea as it relates to a bee's past and the present. Now, I turn to the future, but my reasoning remains similar: bee behavior is too complex to be explained entirely by instinct and by a lack of awareness of the future. While we are likely never to know what it is like to be a particular bee, we can know through careful ethological observations and experimentation that she can have a world in a meaningful sense. In this regard, Chittka describes interesting experiments that demonstrate brain waves ("neural oscillations") in bees, even when they are sleeping. Like that in humans and unlike a machine, a bee's brain is never fully inactive. Furthermore, bees respond to some of the same psychoactive agents as humans, and, like humans, there is suggestive evidence that bees have "place neurons" and "number neurons." Carefully interpreted, reductionist approaches such as neurobiology offer can provide hints at what it is like to be a bee, but this is not Heidegger's approach, nor does it tend to treat bees with the dignity that they are. In the next chapter, I will cite some neurobiological observations, but only within the frame that they enlighten our understanding of other animal *Daseins*.

The future-oriented framework just laid out for what it means to the worker bee to gather nectar can be directly applied to other activities. But first, it is important to emphasize that, while survival of the colony is an important issue, it is not precisely, exclusively, or even principally what I am talking about. The survival issue is the concern of evolutionary biology. Instead, what I am trying to understand is an alien (for humans, but not for bees!) way of living in which such creatures as bees come to experience a world—their world. In gathering nectar, bees both encounter and construct a rich world that includes flowers and their environs. Yes, flowers are sources of food—and perhaps of beauty—but equally significantly they provide the affordances in which the bee's own past and future become manifest to the bee in a present way of living. The flowered environment—including other bees and predators, sunshine and rain, earth and sky—enables being-in-the-world for a bee.

So what are other activities of bees that reflect being-in-the-world? One is the defense that guard bees and workers provide for their colony as a whole against intruders, for example, honey-seeking bears. In doing this, bees are not simply fighting; they are fighting for their colony so as to maintain its way of life. Part of this entails seeing the bear as an enemy, as a threat to their colony. They do not attack the bear randomly but aim for exposed body parts (e.g., face and ears), indicating a knowledge of bear anatomy that is expressed within the context of bringing about a successful counterattack. If successful in leaving her stinger in the bear, the now-partially-abdomenless bee will die. But to say that she dies for her colony is a materialistic underinterpretation; instead, she dies so that her colony may continue to thrive in its way of living. But this is not simply a selfless act. Part of what it means for her to thrive in the colony is that she be ready to die if a threatening situation arises. The possibility of death is a meaningful aspect of her own way of being a defender of her own way of living as a member of the colony, analogous to Heidegger's own characterization in *Being and Time* of death for human *Dasein*.

Up to this point, I have interpreted two seemingly ordinary bee behaviors (searching for food, defending a colony) in ways that illustrate how they are the projection (and projecting) of possibilities. Chittka documents other behaviors (e.g., the workers' and queen's searching for a new hive location after colony fission) that could be similarly interpreted—tasks I leave to the reader. Instead, I want to discuss a new approach—one that is central to Heidegger's understanding of human *Dasein* in *Being and Time*—namely, the role of anxiety in altering one's possibilities. Given that anxiety is a fundamental mood, the "beenness" of *Dasein*'s temporality cannot here be separated from its "future." After discussing anxiety in humans, I ask whether it might also apply to bees.

For Heidegger, anxiety is a fundamental (ontological) mood that is revealed in the form of human *Dasein*'s own being (i.e., its current mode of existence) as under threat. Anxiety reveals *Dasein*'s future as finite, with death as inevitable. But here death refers not to biological demise; rather, it refers to the loss of possibility, or better, the loss of making-possible a future. Anxiety comes upon one because one's own being matters, and oftentimes one is at a loss to explain why—one is just anxious. For example, a society may be on the verge of collapse, but its citizens have learned to adapt to ongoing minor disasters without overtly considering that their life as citizens may end. Even so, they may be anxious without knowing why, in which case the tenuousness of their existence—the possibility of their death—has become their own. At some point, their own anxiety may become an overt realization as societal breakdown amplifies, but still it is not fully comprehensible. In considering this particular example, one could limit one's understanding to the ontical (e.g., to this event or that) but, for Heidegger, to do so would be to miss what it means for a citizen to live in such a situation—to confront (not necessarily overtly) the loss of possibility—and making-possible—that one continues to exist. Of course, this still requires that Heidegger interpret the ontical record.

In thinking about Heidegger's account of anxiety, we might wonder how widespread such anxiety is. Is it to be found among Indigenous nations and other non-Western societies? Was it found prior to the technological era—for example, in the time of Ancient Greece or even earlier? Is it found in other species? Heidegger engages comparative questions only very briefly—and unsatisfactorily—but they are central to my claim of bee *Dasein*. For example, a notable feature of those worker bees who live in temperate environments is that they live just a few weeks, or about 0.1 percent that of a typical American human today. It is hard to believe that this dramatic temporal difference cannot be important in understanding their respective *Daseins*. Might it be that humans living in technological societies may devote much time and effort toward masking or diverting their anxiety as concerns meaningful projections toward death, while worker bees, whose lives are more transient, may remain more focused in their meaningful projections toward death. Might different lengths of lifetime reinforce different forms of *Dasein* and different ways of being anxious toward death?

So let us return to bees and begin with the example of a worker bee who becomes lost while foraging. In this case, "lost" does not simply mean that the bee's navigation system has failed to adequately track her movements, as though a complete explanation resides in this biological reduction. For the bee herself, a navigation system is not primarily about knowing one's location in physical space; rather, it is about extending the range of her at-homeness in order that the patches of flowers become part of that home, even though only temporarily. Describing a bee as "lost" is to say that the bee herself is no longer at home. During an initial indeterminate period, she may fly in ever-broadening circles to see whether she can overlap some familiar sight or smell, but if she fails at this—if she remains lost—then she is no longer at home. Indeed, she is no longer a worker bee as such because this entails that she be at home, that she be part of her colony. What does a lost bee do? Perhaps she never gives up looking for her colony, finally dying fully exhausted. But perhaps she discovers other bees and she follows them. If they return her to her hive, then she is at home once again, but if they take her to another hive, she may be attacked and killed. Either way, she is unlikely to find herself in a situation that is devoid of other bees. Analogously, humans can be lost in an unfamiliar urban space—surrounded by strangers.

Should we consider the worker bee to be anxious only when she is lost? I think not. Rather, she lives in a world where the possibility of existential lostness is real for her. (I am now distinguishing two senses of "lost," an ontical one and an existential one that is underlain by anxiety.) The breakdown condition (i.e., when the bee is ontically lost) reveals to the bee that she has been anxious all along—that her at-homeness and, relatedly, her future are always at stake. In response to this realization, her behavior changes; that is, she starts to search for familiar landmarks to "fix" her loss of orientation, analogous to what the carpenter does when the hammer malfunctions. According to my interpretation, the "busyness" of bees reflects the anxiety of their very existence; it should not be seen as something (simply) mechanical or biochemical but rather a feature that reflects the precariousness of their existence. At times, the busyness of bees is anthropomorphized by self-congratulating middle-class Americans or used as a trope by conservatives to criticize "lazy" people, but human existence itself has been precarious throughout most of its history—and still is for many people. For our fellow bees, beyond the "normal" struggles to thrive—finding enough food, avoiding predators without avoiding the flowers on which they may lurk, sustaining the temperature of the colony in the depth of winter—they are now confronted with massively looming, human-caused stressors, including habitat loss, habitat pollution, climate change, and infectious diseases transmitted from the burgeoning population of domesticated (and abused) animals. And for so-called "domesticated" bees, their struggles are compounded by the frequent, disorienting relocation of the hive boxes to novel environments in order that they provide their "pollination services" for human crops, and by parasite infestations of their boxes. Do these bees not live in a world where the possibility of existential lostness is real?

Breakdown situations and rare situations that affect the entire colony provide opportunities for us to probe a bee's anxiety. For example, Maurice Maeterlinck mentions in a footnote (Maeterlinck 1901) that certain honey bees of Barbados who lived proximate to sugar cane refineries stopped foraging for flowers and instead fed directly on sugar produced by the refineries. If this report is correct, imagine what it must have meant for the bees. No transient fields of flowers to navigate. No predators waiting for them on the flowers. A super-abundant source of food. And yet, a new environment to navigate with new uncertainties. One could say that the bees' behavior is little more than spur-of-the-moment opportunism, but opportunism only operates within the relatively familiar, and the sugar cane refinery is a radically different environment than a patch of flowers. Heidegger says that anxiety may open up the possibility of a new future in a moment of vision. For those cases of human *Dasein* who see these newly presented possibilities as their own and not simply the result of following others, *Dasein* can become "authentic."

In the case of these Barbados bees, they saw a new future that was both imagined and radically different from other bees, and they embraced it. By doing this, they manifested a behavioral flexibility that can be interpreted as opening a new world to their very own authentic *Dasein*.

So far in this chapter, I have tried to make reasonable the claim that a bee can exist in *Dasein* and in a world which is meaningful to her. Of course, *Dasein* must be aware of that world and be able to interpret it as meaningful. To interpret is not to determine, it is to make sense of other entities and of oneself. But, is the world first meaningful and then *Dasein* discovers entities therein, or does *Dasein* first make sense of (i.e., find meaningful) entities such that a world is then disclosed? The answer is: neither and both. It is "neither" in that Heidegger's time is not linear like the question assumes, so the question itself is misconceived. It is "both" in that being is always the being of an entity, so "discovering" and "disclosing" are not separate "events."

The way in which Heidegger writes about individual human *Dasein*—both inauthentic and authentic—makes clear that he rejects psychological assumptions of interiority and autonomy. For example, in describing inauthenticity, he says that human *Dasein* relates to social norms as "the anyone" (*das Man*), a distributive singular term (see Haugeland 2013), for example, a *typical* person. By contrast, in authenticity individual *Dasein* is now his/her ownness, but that "ownness" is not as an observer who is separate from the world but rather as someone already *in* the world. Individual bees—in their *Dasein*—also work to maintain meaning in their (very different) world, most of the time by following others (as for humans), but sometimes by being their ownness, when, for example, they do not attend to the immediate needs of the colony but instead explore the terrain in search of other sources of food, or when they guide the colony in moving to a radically different source of food.

After *Being and Time*, Heidegger also writes about "a people" who work within "a tradition" as being in *Dasein*. Here, "people" does not refer to a particular society-defined collection of individuals, say, the German people of Heidegger's time. Indeed and as described in *Contributions*, Heidegger's "people" may not even yet be. According to Heidegger, what a people is entails deciding—not necessarily knowingly—who we (that is, you and I) are as a people within the frame of (individual) human *Dasein*'s own self in its being-with-others, thereby determining history and the future. Like humans, bees live with others, and not just any others. The ones who matter most are those in their own colony. And as for a human people, a bee colony can also be seen as the resolute projection of bee *Dasein* and as a deciding who bees are in their being, thereby bringing about the colony's history and future. Still, there are differences, and that is why I argue that *Dasein* comes in different forms.

So, what more can we say about bee *Dasein* and existence? While it is obvious that, ontically speaking, we know much more about the human than about the bee, I think we now know enough to interpret the bee's way of being as existing in *Dasein*. In support of this, this chapter has documented that the bee is highly responsive to her environment and—in many cases—in a flexible manner. That the bee responds flexibly to non-natural situations argues against strict rule-following. Furthermore, bees within a colony manifest a range of capabilities, but—like humans—not all bees are the same. For example, some bees are more curious and intelligent than others. A bee's ability to learn argues for her individuality. Bees can also be differently attuned, depending on the situation. Some bee attunements are more fundamental than others, but all bring a bias to the individual bee's behavior. A bee can also have goals (e.g., find new food resources, bring food to the hive, find a new hive location), and such goals—while frequently widespread among members of the colony—can change over the lifetime of a bee or in response to environmental fluctuations. Again, this argues that we ought to interpret the bee as an individual whose way of being, like human *Dasein*'s, is existence.

All this is not to claim that some behaviors cannot be outside a bee's—or a human's—awareness. Bees do display purely instinctual responses. For example, naive bees can construct a hive—a seemingly complex task—with broad features that approximate a normal hive. Even so, (1) we need to overcome the widespread bias that a bee's life is entirely and purely instinct-driven. This mistaken view can be encouraged when—for practical reasons—scientists intentionally bias their own studies and observations toward the study of strongly stereotypical behaviors, and thereby bypass the inherent complexity of other behaviors, or when their experimental aims are deeply reductionist. (2) We need to overcome the bias that all bees within a functional grouping (e.g., workers) of the same colony are the same. Said differently, bees need to be seen as the individuals they are. (3) We need to overcome the anthropocentric bias that occurs when we evaluate bees as such relative to our own capabilities. While Heidegger acknowledged the tremendous ontical diversity of our fellow animals, he was unable to overcome the above three biases. Had he done so, he would have been able to provide a richer interpretation of their way of being.

C4. Spatiality and Bee Dasein

So far, my focus on a bee's way of being has been largely within *Being and Time*'s framework of temporality and its characterization of *Dasein*'s way of being as existence. Later in his career, Heidegger comes to see this framework of temporality, and relatedly that of the "there" (the *Da* of *Dasein*), as insufficient for understanding being (e.g., see *Contributions*: 298). In part, this is because spatiality is considered in *Being and Time* to be a derivative mode of temporality rather than coeval and coequal within a unified time-space. My own understanding of Heidegger is that he does not so much reject his earlier account of spatiality as indicate that it is insufficiently fundamental, or perhaps incomplete—not dissimilar to what he does with his understanding of regional being. In this subsection (C4), I consider what Heidegger says about spatiality in *Being and Time* and in his later works (principally, *Contributions*), and I make propaedeutic efforts to relate these to bee *Dasein*.

Even though—in *Being and Time*—*Dasein* is a "distributive singular term" (i.e., *Dasein* does not refer to a *localized* being), there is still in "*Dasein*" an understanding of self/non-self that makes room for Heidegger's understanding of spatiality. "Spatiality" is a formal indication for the space of human existence, one that Heidegger contrasts with a scientific understanding of space as objects and events in extension (see Wills et al. 2010). Heidegger discusses two instances of spatiality—the nearness and the farness of entities. For example, nearness comes with *Dasein*'s *being amidst* entities, but not in a way that is based on a measure of distance. For example, a hammer can be two-feet distant and still be "near" in a way that the very glasses one is wearing are not. What "nearness" indicates is the possibilization of *Dasein*'s activity (e.g., hammering) in a "place" (e.g., the site of the hammer) within a "region," (e.g., the carpenter's workshop). To think of a place, or space, in isolation from that region—as scientists do—first requires abstracting away *Dasein*'s being-amidst.

Within this context of space, it takes little further imagination to argue that spatiality is central to a bee's existence. Bees are strongly connected to their own hive—to the many places ("nearnesses") within this region—and to other regions, for example, where the gathering of nectar occurs. Indeed, it would be difficult to say what a bee "is" without mentioning place, since place is central to understanding most of a bee's interactions with other entities and, more generally, to what the life of a bee is about.

Heidegger's account of spatiality in *Contributions* is more abstract but can be partially demystified by understanding four key terms: (1) "open," (2) "clearing," (3) "time-space," and (4) "dwelling." (1) "Open" points toward a domain of being in which things (in Heidegger's special ontological sense of that word; Heidegger 2013f) in their truth or essence are accessible; it also points to a particular sort of domain—the open—versus other domains that always remain hidden—early on called "the nothing" and later on, "the earth"; and it points to the very dynamism of opening itself. In *Contributions*, the open is the time-space of decisions, the most inclusive one being whether the human being exists historically—including as a people—in the truth of be-ing or not. Less inclusive "decisions," for example, ones that relate more to an individual *Dasein*'s own existence, are also possible (Polt 2006). The open can be imagined as the very happening of being, as the *place* of this happening, and as contrastive with the unopen. Art—and language generally—are ways in which a thing can be brought into the open for human *Dasein* (see "The Origin of the Work of Art": Heidegger 2002b). We might wonder whether the constructing of a beehive and the interactings surrounding the bee dance are also partially free expressions of the open and of being itself, and not just the utilitarian phenomena that ethologists describe.

(2) "Clearing" is a term closely related to "open." In *Contributions* (p. 275), Heidegger says, "*The essence of truth* is the *clearing for self-concealing.*" The clearing is now the ungrounded ground, or the grounding abyss, or perhaps "a ground that prevails while it *stays away* and, strictly speaking, is the hesitating refusal of ground" (see the translators' forward in *Mindfulness*; Heidegger 2016c: xxi). For a human being to be, she must become open in this clearing, where the struggle

between the comfort and false certainty of her current existence (in the first beginning) and something completely other—namely, the truth of being (in the other beginning)—takes place.

So, does it make sense to talk about bee *Dasein* as in a clearing? Heidegger clearly thinks not. In "Letter on Humanism," Heidegger (1998b: 248) says,

> Because plants and animals are lodged in their respective environments but are never placed freely into the clearing of being which alone is "world," they lack language. But in being denied language they are not thereby suspended worldlessly in their environment. Still, in this word "environment" converges all that is puzzling about living creatures.

Had Heidegger been able to imagine the environment as animal *Dasein*'s analog of human *Dasein*'s workshop, had Heidegger been less anthropocentric in his understanding of language, "living creatures" and "environment" might not have been so puzzling. But failing this, he bypasses serious thinking about bees (and other creatures) and their relationship to their built environment. Instead, he is satisfied to attribute to instinct what individual bees do. But in Heidegger's defense, he is talking ontologically. In *Contributions* (p. 42), Heidegger says,

> The most proper "being" of humans is therefore grounded in a belonging to the truth of being as such, and this is so, again, because the essence of being as such, not the essence of the human being, contains in itself a call to humans, as a call destining them to history.

In his placing "being" in quotes, Heidegger has identified the enigma of talking both about being of the human and of being itself. As in the fourfold, Heidegger can push the human being ("mortal") to the side—calling it a "moment" rather than a "part" of be-ing—but the human still remains necessary and, thus, does not undercut his account in *Being and Time*.

(3) The third key term for understanding Heidegger's usage of spatiality is "time-space." Time-space is not simply a feature of human existence; rather, it stands in the open and into the event of appropriation (adaptation, *Ereignis*), an event that is based on a unwillful "decision" that is separate from, while still dependent on, humanity as seeker of be-ing, preserver of truth, and steward of "the stillness of the passing by of the last god." Time-space is not a presumption or some kind of initial condition, as are time and space in our ordinary way of thinking about human experience; rather, time-space follows from the "decision" of whether to be in the truth of being or not. Time-space grounds the "there" of *Dasein* (literally translated as "being there") as the *site* of the deciding moment of vision (see Table 1.1), but it grounds in the sense that it refuses a ground (and particularly the distorted ground of metaphysical thinking) and thereby opens up an abyssal ground. As an aspect of the clearing, time-space's hiddenness gathers and shelters the truth of being through transporting *Dasein* (Think: timing) and "enchants" in an encircling hold as an ecstatic movement (Think: spacing).

(4) Lastly, let's turn to "dwelling." For Heidegger, space is the site where humans dwell. But in *Being and Time* (e.g., see pp. 88–90), Heidegger's understanding of dwelling is more diminished than later on. In *Being and Time*, "dwelling" refers more straightforwardly to simply *being amidst* a network of "equipment" in one's world, rather than—as later—emphasizing a self's *involvement* while being amidst. Key to understanding his expansion of this term is that Heidegger—in *Contributions*—rejects his earlier account of spatiality as a mode of temporality. Instead, he introduces the term "time-space," and this allows a linkage to be made to the clearing (*Lichtung*) or "the there" (*das Da*) in *Dasein*. While "clearing" points toward the "illuminating" that is being (hence, a dynamic feature of temporality—and time-space), illuminating still demands its proper place (hence, the spatial aspect of time-space),[10] and in that proper place *Dasein* can dwell.

Fortunately, Heidegger eventually published two less abstract essays on space and dwelling—"The Thing" (Heidegger 2013f) and "Building Dwelling Thinking" (Heidegger 2013b). In the latter essay, Heidegger proposes an ontological understanding of world as the onefold of a fourfold—earth and sky, mortals and divinities (discussed in my Chapter 2, Section B). Within this ontological frame, the clearing of being occurs at things, not to be understood as scientific entities but rather (in part) as places or sites (not to be thought in the three- or four-dimensional space of science) which gather the fourfold. Examples of such sites include a suitable work of art, a ceremonial jug, and a local bridge. For bees, they could include the hive and the flower patch. Now, the clearing of being happens as things themselves constitute places or sites, inviting the fourfold and allowing mortals to dwell. The dwelling of mortals is not some additional feature to the clearing of being but is the very happening which invites the fourfold. We could summarize this by saying that from *Being and Time* to "Building Dwelling Thinking," Heidegger shifts from (a) an emphasis on human activities (e.g., that of the carpenter) that let in entities (e.g., various tools within a workshop) that have spatial relations to *Dasein* (e.g., in their nearness and farness) to (b) things that open up spatiality and invite human *Dasein* to dwell (e.g., a path suitable for walking and conversing *or* a pond suitable for fishing). While things are distinct from human *Dasein*'s activity, they are not independent. We might suggest that the things that engage bees, both self-constructed (e.g., the nest) and natural (the flower patch), are also places which invite bees to dwell.

Like being-in[-the-world] as discussed in *Being and Time*, dwelling among things in "Building Dwelling Thinking" should be simultaneously thought temporally and spatially; hence, time-space. Mortals exist only among things, and these things are not just anything but this thing here (spatially) and now (temporally). These "things" allow humans *and bees* to cultivate new things (e.g., the walking path cultivates new human friendships *AND* the bee dance cultivates social cohesion and successful foraging) and to build new things (e.g., the bridge builds a new community on the other side of the river *AND* the flower patch encourages the discovery of additional flower patches).

The later Heidegger moved beyond *Being and Time*'s account of being as temporality by giving due credit to spatiality through his understanding of time-space. In a metaphorical sense, the world became more "real" by allowing a movement from the largely temporal and universalist world imagined by Western philosophers to a world which is necessarily situated, hence, the uniqueness of the event of appropriation (adaptation). Clearly for Heidegger, situatedness relates primordially to the meaning, or truth, or being (or be-ing) that is manifest in dwelling in the fourfold, and only derivatively to the largely meaningless natural world as understood by biologists. But even though understood as derivative, place as an ontical location where meaning can happen in *Dasein* is not a non-understanding but rather a partial or incomplete understanding. That Heidegger failed to understand bees as dwelling in their own space and time-space was his loss.

With all this in mind, let us consider in greater depth the bee's hive as it relates to spatiality. Previously, I described how its construction results from a combination of flexible behaviors and purely instinctual routines, but now I want to emphasize another aspect. The hive is not an edifice devoid of meaning for the bees; rather, it is their dwelling. Speaking ontically, to dwell is to inhabit a world in a meaningful place (i.e., the hive) and time (i.e., during that particular colony's lifetime). But it means more than this—it also means to be "at home" in that place and time. To be at home is a way of being; it is an "activity" as well as a "site." We might, for example, reconsider in a slightly different light the case of a bee who has become lost while searching for nectar, and despite her best efforts remains lost. That bee is no longer "at home," unlike the other bees in her colony who are out collecting nectar far from the hive but who are still "at home."

In the ontological sense, "to dwell" indicates that an underlying "activity" has given rise to possibility itself, including the possibility of existing in bee *Dasein*. We might wonder what this "activity" is, but since it may very well be being itself, we are unlikely to arrive at a direct answer. In *FCM* (p. 144), Heidegger alludes to one manifestation of the "activity" when he uses the phrase "making-possible of [human] Dasein as such" in the context of profound boredom's moment of vision, when new possibilities have been imagined (discussed in my Chapter 1). So by analogy, when bees construct a hive, they are making possible a future where the colony can dwell. Once the hive is complete, bees do not start building elsewhere as though that is what they instinctually do; rather, their behavior changes to meet the changing demands of dwelling together. Completing the construction of the hive opens up new possibilities of dwelling for the colony. But even while building the hive, new possibilities can be imagined, as when, for example, an experimenter's inserted barrier artificially blocks the normal extending growth of the hive.

By contrast to wild bees whose lives mostly revolve around a single hive, human beings who live in highly technological societies sometimes glorify their ability to move into new ontical spaces. Extreme examples of such spaces include outer space, the ocean benthos, and Antarctica. More mundane examples include immersing oneself in a new college life or in a new work environment. We might wonder whether dwelling—rather than non-dwelling—occurs in such spaces. In other words, can technology-focused humans even dwell (ontologically speaking) when one space is as potentially habitable as another? And if we are not dwelling, can we still be in *Dasein*, or *Da-sein*? Or might it be possible to see our dwelling as spatially extended in a way that does not occur for the bee? Perhaps technology-focused humans are in a particular sub-form of human *Dasein*. Vine Deloria Jr. (1973) has forcefully argued in philosophical terms that many Indigenous American nations are deeply and respectfully tied to the particular lands they inhabit. This same point has been made in the poetry of many Indigenous Americans (e.g., Harjo, Howe, and Foerster 2020). A comparison of how these poets relate to their own Indigenous nations with how Hölderlin's poetry relates to the German people (at least as imagined by Heidegger) could be informative but is well beyond my own ability. Even so, the idea that *Dasein* can be place-specific for different groups of humans is consistent with there being multiple sub-forms of human *Dasein*, and correlatively, this makes more understandable the case for the various animal *Dasein*(s).

The later Heidegger's ontological understanding of human spatiality (and temporality) provides a useful interpretative framework for bees, despite the radical otherness of a bee's ontical spaces. For example, a worker bee's relationship to flowers is a central feature of her existence in a way that is largely foreign to humans. Flowers are a place of significance for bee *Dasein*. Still, it might be thought that humans—because they construct many of their own meaningful spaces—must have a richer notion of space embedded in their way of being. But this ignores the fact that bees too build their environment—quite beyond their hive. Indeed, bees—along with other pollinators—are premier global eco-engineers with their pollination services having shaped the evolutionary trajectory of flowering plants which, in turn, have opened up new possibilities for bees (e.g., consider their transition from ancestral carnivory to herbivory). Their eco-engineering also opened up new possibilities for humans (e.g., consider the evolution of our major food crops and its cultural ramifications) and for other life forms (e.g., consider the availability of new foods and ways of shelter). I want to be clear here that my claim applies to *individual* bees, and not just to bees as a group, as scientists are wont to do. The "engineering" resides in the individual bee as who she is, shaping—and accommodating to—her environment.

I conclude this chapter with a cautionary comparative note. My discussion of "bee *Dasein*" has been focused on individual bees within the worker castes of a few closely related species. I have already noted that there are 20,000 other bee species with distinct life histories about which I have made no claim of *Dasein*. But what about bees in other castes of the few species I have already considered? Let us briefly consider the queen bee. The queen's overall behavior is less exploratory than the workers, so social learning may be less necessary. The queen is genetically identical to her workers but has a distinct morphology and behavior due to receiving—from the workers—a special diet. Upon emergence from the pupa, the adult queen first engages in mortal combat with other potential queens, then (if still alive!) departs the hive on a mating flight, and then returns to begin a lifetime of laying eggs, departing her hive—along with a large swarm of workers—only if a new queen is raised in future years. Unlike the short-lived worker, the queen may live for years (Law 2021). There are several points of potential ontological interest here. The newly emerged queen is uniquely *attuned* to searching for, recognizing, and then destroying all other potential queens; she *presences* in exploring the hive and searching for a mate outside; she *projects* in her aim of becoming the long-term and sole queen; and, most of this occurs in and through the *spatiality* of her dwelling in the nearness of her hive and her fellow bees. Interpreting all this as indicating an awareness of her own situation and her situatedness—as I have just done—is a beginning toward claiming her to be in *Dasein*, but my sense is that, unlike the case for the worker bee, more ontical evidence is needed, and particularly experimental evidence, before a claim of queen-bee *Dasein* can be compelling. Even so, interesting issues are already raised: If the queen is in *Dasein*, is her form of *Dasein* (see List 3.1) the same as that of her workers? Or might there be multiple forms in a single colony? Or can we imagine how an entire colony as a colony might be in *Dasein*? And if she is not in *Dasein* (while worker bees are), then we see that in this case *Dasein* does not map to the entire species, even though worker-bee *Dasein* does map across parts of multiple species (namely, the worker castes of the honey bee and bumblebee). All this further emphasizes the importance of identifying and demarcating existentialia in *Dasein*'s various forms. It is an endeavor that will require new ontical and ontological thinking. This endeavor builds on, but is nonidentical with, the systematic endeavor of evolutionary biologists and natural historians, as well as ethologists.

Chapter 5

On Animals: Rethinking Heidegger's Claims

In Chapter 4, I interpreted the ethological record of certain bees within the conceptual framework summarized in List 3.1, namely, that—as *Dasein*—their way of being is existence, but that—like humans—their way of being can also be captivation, in which case they are not in *Dasein*. This plurality of ways of being applies to ontical creatures such as bees and humans, but not to *Dasein*, who although considered ontical has a special status—according to Heidegger—as a bridge (but for humans only) between the ontical and ontological. In this final chapter, I extend my interpretation of the ethological record to a greater diversity of animals, namely: arthropods (other than bees), octopuses and other cephalopods, fish, birds, and mammals (but not including humans). In practical terms, the broad taxonomic scope of this chapter prevents the in-depth of interpretation presented in Chapter 4. Instead, I highlight particular aspects of animal behavior that make reasonable the interpretation that many animals exist in *Dasein*. Heidegger's imposition of an abyssal ontological boundary between humans and our fellow animals was the result of claiming an unwarranted essential specialness for the human while understanding very little about the ontical lives of our fellow animals. Perhaps some animals do not have existence as their way of being, but unless we expand our probative investigation to these animals, what can we claim—and why?

As for Chapter 4, the current chapter continues the interpretation of (ontical) animal behavior within the ontological framework of *Dasein*'s existence as temporality—beenness, presencing, projection—and spatiality. But first, a methodological point: Ethologists document animal behavior by observing in natural and captive settings. Further, these observations occur either with or without experimental intervention. In natural settings, observed behaviors are more likely to be normal behaviors (i.e., not influenced by the human observer), but—practically speaking—observing in natural situations can be challenging and time-consuming. In captive settings, observing may be more convenient, but behaviors may be affected by the artificial surroundings. Experimental interventions—designed to induce behavioral responses that guide hypothesis testing—can be powerfully informative, but sometimes the relevance of a response to the animal's normal behavior may be questioned. When studying domesticated animals, distinctions between (1) natural and captive settings, (2) observations and experiments, and (3) the observer (or experimenter) and the animal subject are less sharp (e.g., see Bastos et al. 2024 for a study of communication between dogs and their human caretakers). These, then, are the methodologies we are presented with in our effort to interpret an animal's way of being.

Section A. Arthropods

At the level of species, the majority of total animal diversity described in the scientific literature resides in the phylum Arthropoda,[1] so if *Dasein* is claimed to be taxonomically widespread, arthropods cannot be ignored. In Chapter 4, I made the case that individual workers in colonies of honey bees and bumble bees exist in *Dasein*, but that they are also in captivation. In this section, I aim to extend that claim to other arthropods. I start by considering the ontological significance of insect migrations, confining my discussion of this large scientific field to a single recent report of a macromoth—the death's-head hawkmoth (*Acherontia atropos*, Sphingidae; see Menz et al. 2022), which is known for startling acoustics and the complex behavior of stealing honey from beehives. *Populations* of these hawkmoths are known to travel nocturnally each spring from their wintering grounds in the Afro-Palearctic region to breeding sites in Europe north of the Alps, a distance of up to 2500 miles. But—the authors ask—do *individual* moths maintain control of their flight trajectories in the presence of head- and cross-winds, or are they relatively passively transported? Fourteen moths were each fitted with small GPS transmitters and individually tracked—eight at higher resolution—over an approximately 40-mile/2.5-hour stretch of their generally north-to-south migration to Africa. Three distinct behavioral strategies were observed: (1) When flying primarily into headwinds, the moths (4 total) took the most direct southward route, adjusting their flight direction as required when the wind direction changed. (2) When flying primarily in the tailwind direction, the moths (2 total) followed a straight southwesterly path that was close to the wind direction but adjusted to be somewhat south of the strongest winds. (3) Also when flying primarily in the tailwind direction, one moth flew directly downwind (west-southwesterly). Median ground speed was similar for moths across the three strategies, indicating self-modulation of powered flight. Further, the consistent straightness of their nocturnal flight paths under variable wind conditions suggests that these moths have some sort of internal compass for maintaining direction. The authors also suggest that these moths use large topographical features—like the mountain range they must traverse—as visual cues to avoid the highest elevations. In conclusion, the authors emphasize that their study of death's-head hawkmoths illustrates that flexible migratory strategies are not confined to vertebrates.

Now let's consider what this might mean for the moths themselves. Foremost is the realization that—during the actual migration—these moths dwell across vast spaces that are frequently changing. From the above report, we see that the moths are not simply passive recipients in the spatial and temporal outlay that is a migratory life. Instead, they are keenly aware sensorily of their changing environment, and they make corresponding adjustments. That is who they are. Unlike we who are relatively sedentary and social, these moths are migratory and relatively solitary.[2] Asking how they know where to fly and what the developmental and evolutionary advantages are of migration are good scientific questions, but they do not address what their way of being means for the moths themselves. Is this a ridiculous question? I think not if we think holistically and non-anthroparchically. Moths have a past that affects them; for example, somehow past knowledge of the species' migratory path is imparted to them. They have a present—as scientific studies are wont to emphasize—but this present can also be seen as a presencing. And they have a future, whether that be fostering the next generation of moths in Europe or recuperating in Africa after a long journey. All three—along with their spatiality—interact to yield their way of being.

Learning, including social learning and culture, is not confined to vertebrates—or even to social species. Consider the fruitfly *Drosophila melanogaster* and an elegant study of its cognitive capacities and dispositions as they affect mating preferences, and how—through social learning—these capacities and dispositions might lead to cultural transmission and maintenance (Danchin et al. 2018). Laboratory-based experiments were undertaken to test each of five criteria deemed necessary for there to be fly culture inherited through social learning: (1) Flies learn by observing other flies [i.e., there is social learning]; (2) social learning occurs across age classes [i.e., social trait preferences can be vertically transmitted—from old to young]; (3) learned preferences are maintained across generations [i.e., preferences can be stored in memory]; (4) learning occurs from copying specific traits [i.e., in this case it was the pinkness or greenness of male flies, and only that, that mattered]; and (5) learning is maintained over generations by reinforcement mechanisms [discussed below]. For brevity, I only focus on the testing of criteria 1 and 5. To test criterion 1, an observer female is first allowed to watch a demonstrator female while the latter chooses to mate with either an artificially colored pink or green male. It was then shown that the original observer female, when subsequently offered the same choice of male phenotypes, preferred the same color of male that the previous demonstrator female had, and that these preferences could be maintained—without further reinforcement—across multiple generations (criterion 4). To test criterion 5, observer females were allowed to observe multiple demonstrator females, some of whom chose pink males and, others, green males. As long as the demonstrators displayed a preference of at least 60 percent (range: 60–100 percent) in favor of one or the other color, the observer females later displayed equally well that same preference. The authors interpret this as indicating a strong bias for social conformity of sexual preference. Further modeling by the authors suggests that large natural populations of flies should be able to maintain such preferences for thousands of generations; that is, their mating behavior becomes a cultural tradition.

So, female fruitfly sexual behavior can be interpreted within the frame of all three aspects of temporality. Beenness is supported by both memory formation and an attunement in the form of social conformity. Presencing is supported by social learning and a sense of one's self (e.g., one's own preferences). Projection is supported by mating, not as an act but as a sense of what one aims for in being the fruitfly one is.

"Agriculture" may seem an unusual topic to raise with reference to arthropods, but if we expand our thinking of agriculture from being a strictly human means of food production[3] to a way of being as an encompassing endeavor of a *Dasein*, then we can think more broadly—analogous to how we can see the *doing* of biology as a form of *Dasein* rather than the *seeing* of biology as a set of abstract theories or practical tools.[4] According to biologists, at least eight lineages of bilaterian animals[5] have independently evolved the practice of agriculture, in addition to humans. Exemplars are: damselfish (there have been multiple independent origins of agriculture within damselfish), a polychaete annelid (worm), the long-spined limpet, the plough snail, the marsh periwinkle (the last three are in distinct superfamilies of the class Gastropoda), termites, ants, and ambrosia beetles (there have been multiple independent origins of agriculture within ambrosia beetles). Agriculture co-occurs with particular overlapping traits of the animal farmers, some of which are: substrate preparation for crops, planting the crops, fertilizing the crops, protecting the crops and pest control, and artificial selection. But "planting," "fertilizing," and "protecting," for example, are not simple observations; rather, they are interpretations of a way of being. What is that way? For example, should we view the farmer as agential or as symbiotic with the crop? But how does "agential" as a psychological concept relate to *Dasein*, and how does "symbiotic" as a biological concept relate to being-in-the-world? And does it matter whether the crop is animal (e.g., as for homopteran-tending ants), plant (e.g., as for the *Philidris* ant), or fungus (e.g., as for attine ants)? And is it always clear who is the farmer and who is the crop? These questions require nuanced and detailed responses, but my aim is different—to interpret animal farmers as being in *Dasein*. The scientific language of "agency" and "mutualism" is not Heidegger's (or mine), and although it is possible to provide a Heideggerian reinterpretation of these and other scientific terms (e.g., "consciousness"), that is not my chosen task.

To further delimit my response, I only comment on the suite of behaviors that have led to the identification of agriculture in leaf-cutting, fungal-farming ant colonies of the genus *Atta* (see Schultz 2020, 2021). *Atta* originated about twenty million years ago when it diverged from other attine ants who display somewhat distinct forms of agriculture. An *Atta* colony begins when a daughter queen departs her natal nest, is fertilized by multiple genetically distinct males, excavates the beginnings of a new underground nest, lays and tends the first brood of fertilized eggs, and—still within the nest—begins a "garden" with an inoculum of fungi that she has carried from her natal nest. These founding nest activities are performed only by herself—using only energy from her own fat and muscle stores—but they are relinquished once the first generation of workers emerges. After that, her main activity over the next eight to fifteen years is egg laying, as the colony size may grow to five million individuals. Fertilized eggs are always female, and they become workers of several distinct castes—gardener-nurses, within-nest generalists, forager-excavators, and defenders (ordered from smallest to largest).

More than thirty tasks have been identified. To illustrate, leaf fragments are not just cut but also masticated into a paste, the paste is always added to a new area of the fungal garden—which is then inoculated with a fresh portion of fungus. The fungus is harvested and fed to larvae. Workers are able to recognize when particular sources of food adversely affect the fungi, and they then find new sources, while remembering for multiple weeks to avoid the old sources. The garden is actively monitored, and unwanted bacteria and fungi are physically removed or treated with an antibiotic produced by commensal bacteria inside the ant (Pathak, Kett, and Marsavi 2019). Exhausted portions of the garden are actively removed to outside the nest or to special locations within. Although tasks appear size optimal to a particular caste, if one size of ant is experimentally removed, other sizes of ants take over the task. The ant/fungal relationship is a two-way obligatory mutualism, with both partners showing adaptations to the other. The nest is also an abode for hundreds of species of other protostomes (but also snakes and lizards), including obligate commensals.

If we stop thinking about ants of the genus *Atta* simply as little chemical-sensing automatons and, instead, consider their many complex behaviors—not only the tasks outlined above but also their coordinated interactions with nestmates and others that give rise to the "mini-city" that scientists reductively call "a nest"—then, we open up the possibility of seeing them as the ants they are rather than as fungible, property-filled entities useful as a devivified model system for the general scientific understanding of symbiosis.[6] These ants display behavioral flexibility, indicating an awareness of their world. How do we know that they are aware? Is this a strictly scientific question? What we can observe is that individual workers alter their tasks depending on the needs of the hive (consistent with presencing), they are aware of the health of their fungal garden (consistent with presencing), they remediate the garden appropriately (consistent with projection), they remember the effect of a particular food source on the garden's health (consistent with an attunement of beenness), they know how to cull the less productive portion of the garden (consistent with presencing) in order to avoid unwanted infestations (consistent with projection), and they are on the lookout for new sources of food (consistent with projection). Given the large number of tasks/distinct behaviors of worker ants, the hypothesis of social learning (i.e., learning by watching and mimicking others) seems eminently reasonable.

Now considering the queen, her complex behavior from the time she leaves her natal nest to when she becomes primarily an egg layer defies a completely generalizable description because there are queen-to-queen differences in behavior that matter to each queen and to the future of her own colony. If we consider the queen to be in *Dasein*, then this mattering is not to be understood as various internal and external forces acting on her but as an attunement of her own way of being. But, is not her attunement a fixed feature, one determined by the special food she receives as a larva? This way of thinking—once again—substitutes mechanism and emotions for *Dasein* and way of being, or, alternatively, substitutes a reductionist understanding for a (more) holistic one. A queen's attunement and her projection of the possibility of founding a colony—despite the alternative possibility of failure to do this—mesh in a presencing that is ontically manifest in her orchestrated behaviors.

In all this, place matters, where "place" serves as an indication of *Dasein*'s spatiality. Like humans, these ants are born into an obligatory social context. Obviously, this includes other individual ants who become one's ingroup—while also keenly being aware of outgroup individuals—but it also includes nearby environmental structures (e.g., the physical nest, the surrounding geological and biological features) and faraway structures (e.g., "foreign" nests, uninviting geological features). But nearness and farness are not absolute distances. For example, a leaf-harvesting area may be absolutely farther from a worker's home than a foreign colony's nest, but still nearer to her in *Dasein*.

Lastly, social ants—among other hymenopteran insects—have been interpreted as superorganisms; that is, Darwinian individuality is at least partially explainable at the level of the colony (Hölldobler and Wilson 2009). Currently, I am skeptical that this applies to my own Heideggerian understanding of the individual. However, we might also wonder whether symbioses themselves point to a higher-level understanding of the individual, for example, between ant and fungus. Although I find this idea at least worthy of consideration as a biologist, it does not make much sense to me when I try to think like Heidegger. More generally, we might ask: where must holistic thinking end if it is still to allow interpretation?

Now, I turn to questions surrounding an awareness of sensory injury (= pain),[7] a topic of widespread interest from the viewpoint of human relations with our fellow animals. Interestingly, Heidegger himself does not engage with human or animal sensory injury. In *FCM* (p. 271), he does say that an animal's poverty in world may be a "pain and suffering," but here pain is existential rather than sensory, and even this pain is meaningful as pain only from a human perspective. Later in his career, he rails against industrial ("factory") farming, although its relationship to ontical animal pain and suffering remains undeveloped. Rather, for Heidegger factory farming reflects on the negative aspect of letting-be (*Gelassenheit*) for human *Dasein* in its truth and world (see Introduction and Foltz 1993).

But Heidegger's lack of interest aside, and while still following his overall project of *Dasein* thinking, I see awareness of injury as built on the attunement of an animal's *Dasein* to possible aversive situations. This attunement leads to presencing in a world that matters, and as projecting possible futures. This is not how many others describe the topic. Experiencing sensory injury is oftentimes considered to be a private experience, and yet few humans doubt that all fellow humans can be in pain. But on the other side of the linguistic/non-linguistic boundary, people *ponder* whether this animal or that one can be in pain, typically concluding by saying that one can never be sure. And also frequently, animal responses to sensory injury are interpreted as reflexes with no awareness of injury and, hence, no pain. Fortunately, paying careful attention to ethological studies can jar our thinking.[8]

Consider the shrimp who performs prolonged wound-tending of her experimentally irritated antenna but not of the other (unirritated) antenna. Awareness of pain can also be manifest in less direct ways, including avoidance learning, protective motor reactions, and motivational trade-offs (Elwood 2012). For example, some hermit crabs—which use empty shells of marine snails as their primary protection and housing—have been observed to place sea anemones around their shells as (presumably) additional protection against predators, such as octopuses, who avoid the anemone's stinging nematocysts. But the most convincing of Elwood's criteria for an awareness of injury is that of motivational trade-offs, which I now document in the shore crab, *Carcinus maenas* (Magee and Elwood 2013).

Living in rock pools and shallow waters, shore crabs typically prefer less illuminated areas, likely so as to minimize predation, for example, by watchful birds. Taking advantage of this aversion to illuminated spaces, the experimenters constructed a brightly illuminated test chamber in their laboratory that included two well separated and dark sub-chambers. Each sub-chamber had a distinctive visual cue located on its outside. One of the sub-chambers was wired to deliver mild shocks to any crab who ventured inside. Here is a summary of the experiments and four key observations: (1) Experimentally naive individual crabs who were released into the main chamber showed no (initial) preference[9] in their selection of sub-chamber (= trial 1). It is important to emphasize that it was the experimenters who chose which sub-chamber would deliver a shock. With no changes to the experimental set-up and protocol—including which sub-chamber conferred shocks—and with the same crabs tested in trial 1, the trial was then repeated 9 times (= trials 2–10) to assess how, if at all, they responded to the shock, or its absence. (2) Also during trial 1, many shocked crabs—but no unshocked crabs—exited their sub-chamber into the lighted arena, with the frequency of such exits increasing in later trials. Most crabs who exited remained in the illuminated arena for the remainder of the test—thus, overcoming their aversion to the brightness—but of those who did reenter a sub-chamber, a large majority selected the other (no-shock) sub-chamber. (3) Those who received a shock in trial 2 (= their first actual learning opportunity) were more likely to select the other (no-shock) sub-chamber in trial 3 than those who selected the no-shock sub-chamber in trial 2, independent of whether they were shocked in trial 1. (4) At the end of trial 10, the two sub-chambers' distinctive visual cues were switched (= trial 11A) to determine whether this would alter the same crabs' (i.e., those who had completed tests 1–10) choice of sub-chamber. Switching the visual cue was without effect. In an alternative test (= test 11B), the experimenters placed the same crabs in the chamber in an altered orientation without switching the visual cue. Now, the crabs' preference for sub-chamber was reversed; they moved to the shockable sub-chamber.

What do we learn from these observations?[10] Trial 1 is a control in that (1) it demonstrates that the crabs had no sub-chamber preference prior to receiving a shock and (2) it demonstrates that the shock is indeed aversive to the crabs and leads to a change in their behavior; that is, they exit the sub-chamber only if they receive a shock. Trials 2–10 provide the experimental observations: They demonstrate that the crabs both *remember* the shock and *learn to discriminate* between sub-chambers in order to avoid the shock. These observations are consistent with the claim that crabs' have an awareness of aversive sensory injury, that is, with having pain. Performed in primates, these same results would have generated no doubters as to the veracity of that claim. Trial 11 shows that what the crabs use to guide their behavioral change is not the relative location of the sub-chambers but their own directionality in movement. This indicates that—in these crabs—there may be no need to invoke their having a cognitive map of the chamber; rather, they learn by associating shock avoidance with a change in their directional movement. In Chapter 4, I addressed those who question whether learning by association indicates awareness, and I further note that this skepticism likely also extends to "learning" and "memory." These doubts are rooted in biases for mechanistic and mammal-centric thinking. Without reprising the arguments, I note that the words in quotations are used by the co-authors of this report, who were bona fide biologists at the time of these studies.

So far, I have interpreted behaviors in insects (moths, fruit flies, ants) and other pancrustaceans (shrimp, crabs). I conclude this section on arthropods by considering the maternal behavior of an ant-mimicking jumping spider (Euchelicerta: Arachnida: Araneae: Salticidae: *Toxeus magnus*; Chen et al. 2018). This spider, like the tsetse fly (Wei-Chun and Denlinger 1974), provisions her young with a self-manufactured, nutritious liquid secretion (called "milk") for about thirty-seven days. Initially, the mother deposits milk droplets on the nest's inner surface, where the spiderlings learn to feed. After the first week, external milk deposition stops and the spiderlings then feed directly from the mother's own body, that is, her epigastric furrow. After forty days, milk feeding ceases, and thereafter all spiderlings leave the nest to forage, although some start foraging by day twenty. Despite foraging outside the nest, the young still return up to day sixty, although male offspring are not allowed to return after reaching sexual maturity at about fifty-two days—indeed, they are attacked by their mother if they attempt this. Throughout the entire period, the mother maintains the nest by repairing damage and removing her youngs' shed cuticular skins. The latter behavior likely reduces parasite load. The coauthors' experiments demonstrate that the presence of the mother increases the health and survivorship of the young. For example, if she is (experimentally) removed at day twenty, fewer spiderlings reach sexual maturity and full body size. By contrast, if her milk feeding is (experimentally) blocked after day 20 but the mother is still present and attending to her other tasks, there is no statistically significant effect on spiderling survivorship to adulthood or on body size.

What are we to make of these diverse behaviors? Biologists interpret them within the framework of a life history strategy (i.e., prolonged parental care such as found in long-lived vertebrates) that allows the young additional time to mature, thereby reducing predation and disease at the most vulnerable stage and allowing time for skills to be learned, for example, foraging when one's survival does not depend on complete and immediate independence. As for all the arthropods I have discussed, I am not here claiming insight into what it is like to be a jumping spider. What I am claiming is that their various behaviors, when considered together, can be interpreted within the framework of memory and bias, awareness and intelligence and learning, and goal-directedness, and that all of these are focused around a meaningful (for the spiders) space called the "nest." In other words, these spiders can be interpreted as being in *Dasein*. Claims of behavioral instinct are beside the point. Claims that their behavior should be understood as mere mechanism are wrong. In saying this, I am not claiming that their behaviors are too complex to be understood using reductionist or algorithmic languages, but that such understandings remain highly speculative and are premised on an ontological claim—a claim that does not see the spider as a spider. That is what is wrong.

Section B. Octopuses and Other Cephalopods; Dasein's Forms and Biological Research

In considering the beenness of temporality in cephalopod *Dasein*,[11] we look to their affects, attunements, behavioral biases, and memory formation. However, because most octopuses (and other cephalopods) are largely asocial and solitary and are restricted to marine or brackish waters, they are difficult subjects for controlled behavioral studies. Furthermore, given the relative paucity of behavioral observations for any one species of cephalopod, my discussion will highlight behaviors from widely divergent species (see Staaf 2023 for an accessible overview of cephalopod diversity).

Unlike philosophers who emphasize abstract thinking in order to explore one's own world and that of others (e.g., consider Heidegger while he was writing *Being and Time* or *FCM*), octopuses first and foremost explore with their camera-like eyes and their highly sensate and muscular arms. The broadly exploratory behavior displayed by observing an octopus's arms in particular points to a most noticeable manifestation of an octopus's curiosity about entities in their world, for example, a nearby crab or a human diver's novel equipment. We can say that touch-based curiosity-in-motion is an attunement and an existentiale of an individual octopus's *Dasein*. Of note is that control (biologically speaking, of course) of individual arm motion is partially localized to neurons within the arms themselves—neurons that are separate from the brain in the distant head region. But such "semi-autonomous" behavior does not undercut octopus *Dasein*'s unity because (partially) separate "control" by discrete parts of the organism is a scientific reduction of the octopus's own way of being curious-in-motion. Furthermore, octopus eyes are located in the head—even though the arms too are light sensitive and can respond to light—and they "direct" the organism to the object which the arms then further "explore." As compelling a description as arm-versus-head talk is, this scientific way of talking does not address *Dasein* and what it means to be the octopus that it is. Curiosity-in-motion is in octopus *Dasein* and is never confined to a body part. Similarly, talk of the octopus having multiple selves is an unwarranted psychological reduction (e.g., Godfrey-Smith 2016), at least as relates to *Dasein*-thinking. While possibly useful for furthering scientific understanding, talk of multiple selves may impede understanding of octopus *Dasein*.

Octopus curiosity-in-motion is reflected in a willingness to do novel—even artificial—tasks, and to sustain that curiosity-in-motion to completion. Godfrey-Smith (2020) says that octopus behavior can have an "improvised, opportunistic look about it," when, for example, he describes an incident in which a "wild" octopus attempts to camouflage it*self* from human divers (i.e., "the other") by picking up a nearby sponge and hiding behind it. And, in laboratory experiments, octopuses can open jars and solve maze problems, even without seeing where their arms are going, in ways that defy reduction to trial-and-error (Amodio et al. 2019). We can say that individual octopuses are both attuned and clever. Of course, their attunement cannot be cleanly separated from their cleverness and sense of self/other, both aspects of which are central to the presencing of octopus *Dasein* (discussed below).

Octopuses are predators—of shelled creatures (e.g., crabs, scallops), fish, and even other octopuses—and, as such, manifest controlled aggression. Controlled aggression is not something that can be directly observed; it is an interpretation of behavior—neither a thing nor a process—and can be considered another attunement and existentiale of octopus *Dasein*. Controlled aggression affects more than hunting behavior. For example, under certain environmental conditions (e.g., abyssal hydrothermal springs [Barry et al. 2023], abundant food [see Scheel et al. 2018]), octopuses may choose to live in proximity to each other, and then controlled aggression may be expressed through dominance hierarchy and nest protection (see Godfrey-Smith 2016). Also, projectile use—previously documented only in social mammals—has been observed in one congregation of octopuses in the form of throwing shells, silt, and algae. To do this, an octopus picks up the material from the sea floor with its arms and then drops it while simultaneously directing a powerful jet of seawater—using its siphon—toward the material. Some throws were directed toward and hit other octopuses, and these likely indicate aggressive behavior within a social context.

As predators no longer evolutionarily constrained by a shell (N.B., cephalopods evolved from a shelled molluscan ancestor), octopuses are highly mobile and utilize this enhanced mobility in their active search for prey; they are not ambush predators who wait for their prey to come to them. But, with shell-lessness and exploratory behavior, octopuses have become more vulnerable to predation. Enhanced intelligence may have been one evolutionary response—to be further considered when I discuss octopus presencing. Another may have been their relatively short life. Still another may have been their amazing capability for rapid changes in skin coloration that match their immediate environmental background. But skin color can do more than camouflage. For example, aggressive behavior can be expressed toward other octopuses through darker skin coloration, as can particular postures (e.g., one such easy-to-imagine aggressive posture is called the "Nosferatu")(Scheel 2023). Both can be interpreted as interpersonal signaling of one's own controlled aggressive attunement, now activated in a particular manner. As an aside, skin color changes have been observed during a sleep-like state in a cuttlefish (*Sepia officinalis*), and one interpretation is that these express moment-to-moment changes in brain activity (Iglesias et al. 2019).

Given the radical otherness of an octopus's life, some people might still question whether octopus attunements have any connection to those of humans. But here, biology provides some justification in that certain human psychotropic drugs (e.g., the street drug called "ecstasy") also affect the behavior of an octopus (*Octopus bimaculoides*), and in some similar ways (e.g., enhanced prosocial behavior) (Edsinger and Dölen 2018), making reasonable the claim of some shared human *and* octopus attunements.

Now, I turn to the role that memory plays in octopus *Dasein*'s beenness. Attunements would be inconsequential—indeed, they would be completely missing—if the organism had no memory of its biases through which attunements could be expressed. An octopus recognizes and remembers that other individual octopuses have reappeared, that sharks are their predators, that scallops are their food, and that kelp can be something to play with. When in human captivity, octopuses can also recognize and remember *individual* humans and can express and remember their likes or dislikes of and toward them (see Godfrey-Smith 2016). Needless to say, octopuses can also remember locations of good food sources and habitation, and they return to them (i.e., they remember them) when they are hungry or seeking protection (Mather 2008).

Memory has also been examined under laboratory conditions in another group of cephalopods, namely, cuttlefish (*Sepia pharaonis*; Kuo and Chiao 2020). Cuttlefish will choose an offering of two shrimp more frequently than one when both rewards are simultaneously presented. However, when primed with a choice of one shrimp or no shrimp and then presented with a choice of two shrimp or one, they chose one shrimp more frequently than those primed to choose two or one. The authors interpret this to indicate that the cuttlefish remembers its previous offerings in making this valuation. Cuttlefish can also learn and form long-term memories to avoid food that is presented behind a glass barrier—hence, inaccessible (Bowers et al. 2020).

We now turn to the presencing of temporality in octopus *Dasein*. Most generally, this is manifest through an animal's awareness of its world. My discussion focuses on three ways of being aware—awareness of pain, awareness of self, and awareness of the environment, including other living beings.

Octopuses experience pain as evidenced by their behaviors of escape, wound tending, and protection. For example, injury to the arm of the algae octopus (*Abdopus aculeatus*) results in inking and jetting to escape the source of injury (Scheel 2023). Subsequently, the octopus grooms the wound and holds the injured arm in its beak.[12] Later, the octopus contracts the injured arm and—in a protective move—surrounds it with the other arms for a day or more. Furthermore, when lidocaine—a local anesthetic for humans—is applied to the octopus arm, a subsequent injury (namely, pinching) blocks the octopus's normal protective responses. Should we see these striking similarities with ourselves as awareness, or should we turn to psychobabble and question whether octopuses are *truly* aware, where here "truly" means: in the human mode? Heidegger's abyss is another way of raising this same question, but if we take our increasing understanding of octopus behavior as relevant to our ontological interpretation, then, the time has come to admit that the abyss is no longer warranted, if in fact it ever was.

Arm-injury observations and experiments provide further insight, namely, that octopuses (*Octopus vulgaris*) have an understanding of self. For example, when presented with the detached arm of another octopus, the intact octopus treats it as a food item—attaching the suckers of its own arm to the item and bringing it to its mouth (Nesher et al. 2014; Scheel 2023). But when the detached arm is its own, repeated contact is made but the suckers seldom attach. The most immediate mechanism of this self-recognition involves skin, but of course this does not mean that such recognition is merely mechanism, unless one is willing to say that similar human behaviors are also to be considered merely mechanism until proven otherwise. Relatedly, whether the neural connections that are involved in this self-recognition reside exclusively in the arms or whether the brain also provides input—while interesting from a scientific viewpoint—is not necessarily relevant when thinking about octopus *Dasein*. I think the best current explanation is that octopuses have a sense of self. Here, I am speaking both ontically (e.g., naturalistically) and ontologically (i.e., in terms of the octopus's way of being).

I have already pointed out that octopuses are highly exploratory, but if they are going to learn from their explorations, then a precondition is that they be able to re-identify particular objects as the same when viewed from different vantage points (and at different times). Such perceptual constancy has been experimentally demonstrated for the octopus (*Octopus vulgaris*), for example, in its ability to distinguish small and large squares independent of their relative distances (see Godfrey-Smith 2016: 99). With this in mind, we can turn to demonstrations of their ability to learn.

Many octopuses eat crabs, and this requires that they know how to deal with—to manipulate—the crab's protective shell in order to secure a meal. Of course, crabs come in different sizes, and then there are other shelled creatures that they may eat—like shrimp, snails, and scallops—all of which have quite distinctive shell structures. Manipulation of the prey and extraction of the edible portion requires dexterity applied in a context-sensitive manner. For example, when striking crabs, octopuses (*Octopus bimaculoides*) tend to use synchronous arm movement, while for shrimp, they use a sequential motion (Bidel, Bennett, and Wardill 2022; Chang and Hale 2023). As observed under laboratory conditions, octopuses can learn to eat novel shelled creatures (Scheel 2023: 77). For example, when presented with a clam for the first time, octopuses may use the sharp radula located on their tongue to drill a hole, or their beak to chip away at the shell, before figuring out that the two halves of the clam shell can be more easily pulled apart using the suckers on their arms. Additionally, Yasumuro and Ikeda (2018) show under controlled laboratory conditions that cuttlefish (*Sepia pharaonis*) who grow up in an "enriched environment" learn to complete their assigned task at all developmental stages tested and to retain the memory to do this longer relative to those who grow up in "isolated and poor environments," and that this leads to their becoming more capable predators. To carry out these sorts of tasks, we can say that octopuses (and other cephalopods) call upon their various abilities as the situation arises, and that they are aware of their position within a larger space so as to give context-appropriate responses.

As predators but also prey, octopuses are keenly aware of their environment and its possibilities. Using their siphon, some octopuses will squirt water in order to ward off scavenger fish, burrow into the sand on the ocean floor in order to secure a hiding spot while hunting, and fastidiously remove food remnants from their den. They also barricade their den with rocks and coral in order to secure that abode. Such behaviors are unlikely to be simply aversively reflexive because squirting can also be directed toward items that it plays with (see Amodio et al. 2019). Octopuses have even been observed hunting cooperatively with fish, together flushing out prey from places of hiding and "taking turns" eating the prey (see Staaf 2023: 35; Sampaio et al. 2024). Interpreting octopus behavior stretches the human imagination as we try to consider what it means for octopuses themselves to be aware.

Iconically, octopuses (and other cephalopods) can change their overall skin coloration and patterning in order to avoid detection by their predators. We skeptical readers might think skin coloration to be a reflex, but reality is more complex and interesting. For example, a male cuttlefish can have an (honest) courtship display of skin coloration on the surface facing a female while simultaneously displaying a distinct and deceptive female coloration on another surface facing a rival male. Can such behavioral flexibility (i.e., simultaneously transmitting contradictory signals) simply be instinct? Presented with a local-environment-contrasting probe, camouflaged cuttlefish (*Sepia officinalis*) come to break their color pattern more quickly and more consistently if they receive a food reward shortly thereafter rather than at random times (Hough, Case, and Boal 2016). The authors conclude that this cuttlefish response is based on their own awareness of the relationship between breaking camouflage and later receiving food, a response the authors say cannot be entirely reflexive. Morse and Huffard (2022) document another example of contextualized octopus (*Hapaloclaena maculosa*) behavior when they observe that males are more likely to retreat from females with whom they have already mated than those with whom they have not.

Now, we turn to ontological projection in *Dasein* as a framework to interpret octopus behavior. In humans, flexible tool use is typically interpreted as goal directedness, or evidence of planning, and—in a Heideggerian framework—as a specific ontical manifestation of (ontological) projection, or a for-sake-of-itself. Octopuses, too, can engage in flexible tool use. Previously, I indicated that octopuses use their siphon in varied and complex ways. In some instances (e.g., burrowing into the sand), this has been interpreted as using water as a tool (Amodio et al. 2019). Relatedly, octopuses barricade their entryways with rocks and coral, and this too has been interpreted as tool use. But what is a tool? Finn, Tregenza, and Norman (2009) usefully say that a tool provides no benefit until it is used, and that even when used, the benefit derived from tool use is delayed. The octopus examples fit this definition. In a striking example of *composite* tool use, they describe how some veined octopuses (*Amphioctopus marginatus*)—also called "coconut octopuses"—provide for their own (delayed) protection by cumbersomely transporting with them two halves of coconut shells into which they can neatly fit themselves whenever and wherever the need arises. Of course, this requires that the octopus figure out how to reassemble the two transported halves so that they form a protective spheroid. Carrying the shells as such provides no immediate benefit; indeed, it reduces locomotory efficiency and exposes the head and mantle to predators, but this activity must be outweighed by the delayed benefit of having a quickly assembled and secure refuge/abode. This resource (i.e., empty coconut shells) occurs only near human habitation since humans are the source of the shells, but elsewhere these octopuses turn to the use of other shells, for example, of gastropods. In all this, octopuses display their intelligence through curiosity, opportunism, and behavioral flexibility.

Up to this point, I have interpreted cephalopod behavior within the framework of *Dasein*'s three aspects of temporality, but as the later Heidegger makes clear about human *Dasein*, temporality as being is not to be sharply distinguished from spatiality; hence, his formal indication of time-space, or the "site of the moment" (*Augenblicksstätte*; see *Contributions*). Here, "site" points to space, while "moment" points toward temporality, but neither should be thought naturalistically. *Dasein*'s site is how it dwells in a particular situation; it is a way of being at-home in a world. Since Heidegger thinks that modern humans no longer dwell essentially—a consequence of modern technology's overpowering essence that distorts dwelling—we see that essential dwelling is not a necessary condition of the human. But what about the octopus? First, a bit of humility is warranted. Given the paucity of ethological studies, we know very little about what spatiality may be like for a creature so different from us. To begin, let us return to the veined octopus (*Amphioctopus marginatus*). The empty coconut shells in coastal waters are discarded human products that the veined octopus has relatively recently learned to utilize, presumably shifting from utilizing empty bivalve shells, but in doing this, the octopus's world has changed, albeit in small ways. Through the coconut shell, the octopus has formed a connection to human *Dasein*. I like to think that the octopus with his/her coconut shells can still continue to do what they normally do; that is, they can continue to dwell, albeit somewhat differently.[13] But here I need to emphasize that "normally" does not mean "on average" or "typically." Rather, it is meant to point toward an essential way of being an individual octopus.

But, what if discarded coconut shells were to become so abundant as to literally crowd out living marine organisms—like shrimp, snails, and scallops—on which the octopus feeds? Hopefully, the octopus would find food elsewhere and return to its former way of using empty bivalve shells for protection. But what if humans destroyed a large portion of the nearby seafloor habitat (e.g., by large-scale, mechanized scraping of the seafloor for the purpose of capturing bottom-dwelling fish, crustaceans, or minerals)? In other words, could human technology make octopus dwelling impossible in its essence just like it has for humans? And here, dwelling is not simply an ability of the octopus to subsist, although that is clearly essential, but also the continuance of the very doing of one's life activities within the context of a world that is meaningful to the octopus. Has this devastation already begun? Staaf (2023: 92) poignantly shows a photograph of a veined octopus inside a broken glass bottle—a potentially hazardous situation for such a soft-bodied creature. Also, I note that human consumption of wild-caught octopus and squid is increasing, and that cephalopods are increasingly the targets of invasive scientific investigation, neither of which can possibly foster their essential dwelling.

I began this section by saying that octopuses are largely solitary and asocial, but later, I said that they sometimes choose to live in close proximity to each other and to deal with each other. What does this say about how octopuses dwell? It says that dwelling cannot be reduced to a naturalistic understanding of habitat or organism. It says that dwelling is something that the octopus actively does as it surveys its habitat. Through its exploratory curiosity, the octopus engages and understands opportunities, for example, empty coconut shells as homes. In this context, let us now consider Octopolis, a human-diver-accessible site off the coast of Australia where Common Sydney Octopuses (*Octopus tetricus*) congregate (Godfrey-Smith 2016; Scheel et al. 2018; Scheel 2023) due to abundant food and protective habitat. From a diver's initial perspective, this site can be described as a large midden of empty scallop shells into which octopuses have constructed their dens, but at that site many interesting octopus behaviors have been observed. As per Scheel et al. (2018), we might ask which features of individual octopuses at Octopolis are nonintentional outcomes and which are intentional outcomes. As an example, the authors identify the proximity of dens to each other as a non-intentional outcome of food availability and predator avoidance. But might not the ability—or willingness, in psychological terms—of individual octopuses to tolerate each other also be essential for understanding how such proximity could even be possible? Unlike billiard balls, octopuses do not come to be side-by-side by chance; they must also know how to live side by side. Ethological observations allow us to interpret dwelling via the articulation apparent in an octopus's form of *Dasein*, but this interpretation is not forced on us.

Scheel et al. (2018) devote their report's last paragraph to refuting the claim in the popular press that Octopolis is a city, where now the human form of intentionality (e.g., expressed through commerce, culture, and design) becomes central. But surely, the origin and later development of human-cities emerge in part from nonintentional behaviors also, as for Octopolis. Rather than arguing that Octopolis is—or is not—a city, I'm suggesting that "city"—like all words—can be used either broadly to include numerous animal species or narrowly to include but one, and the choice cannot be based on merely looking at some object. Arguing that Octopolis cannot be considered a city in any sense and that it has no connection to a human-city—something that Scheel et al. 2018 do *not* do—is another way to foster Heidegger's abyssal thinking.

Despite its apparent unusualness for the octopus, the group living at Octopolis offers the ethologist important opportunities to assess an octopus's proclivities, capabilities, purposes, and sense of space. At Octopolis, octopuses are continually interacting and can even display social stratification (e.g., a single male gatekeeper of the entire site has been observed), but what does this mean for the octopus? At Octopolis, some new octopuses arrive seasonally, given their relatively short lifespan, but how do these naive octopuses learn the ways to behave in this unusual environment, including continually interacting with others? At Octopolis, who and what are the octopuses watching and signaling? And, how does their reported ability to recognize other individual octopuses fit into this (Tricarico et al. 2011)? Like the arthropods I have discussed, octopuses challenge us to think more expansively about *Dasein*, and they thereby draw us away from our own anthropocentric understandings.

Dasein's Forms and Biological Research

So far, my discussion of animal *Dasein* has been confined to individual protostomes who have been sparsely sampled from two taxonomic groups—arthropods and a subgroup of mollusks, the cephalopods. I have yet to discuss the vertebrates (see sections C–E), and I will not discuss individuals from other protostome groups, including annelids, nematodes, flatworms, and numerous smaller groups. Fortunately for the length of this book, my aim is *not* to be encyclopedic; rather, it is (1) to undercut Heidegger's abyssal claim that *Dasein* is to be found only in the human, but also (2) to undercut any move by others to restrict *Dasein* to a subgroup of, say, deuterostomes (e.g., great apes, primates, mammals, vertebrates) prior to undertaking a deep ethological exploration. But already, my interpretations of bees (and other arthropods) and cephalopods must be jarring to those who, like Heidegger, consider *Dasein* and human *Dasein* to be equivalent terms, and my interpretations of fish, birds, and (nonhuman) mammals will continue this interpretative expansion of *Dasein*.

Having introduced some diverse forms of *Dasein*, it is time to ask again (see List 3.1 and Section B of Chapter 3 generally): How are *Dasein*'s various forms to be articulated with respect to science-based taxonomic categories? Here, there are two large and interrelated tasks, both of which remain mostly incomplete. One—a traditional scientific endeavor—is to identify relevant characters—particularly morphological and behavioral ones—that make possible the way of life for each and every species. The other is to articulate *Dasein*'s multiple forms and their existentialia that make possible a particular way of being (i.e., existence). This latter undertaking must be informed by science, but "*Dasein's* forms" and "existentialia" are not scientific concepts. Still, we must inquire into what ontical diversity a particular "form" actually captures. Properly undertaken and understood, such a *Dasein*-focused task could constitute a new approach to biological research.

I wish to very briefly expand on this thought by way of asking questions. (1) Are existentialia the same across all forms of *Dasein*, or do they vary while still maintaining their link to existence? A reasonable starting position would be to postulate that there is a diversity of existentialia across various organisms, but with some (necessary) overlap across forms. (2) How can we relate existentialia to biological characters? Here, I think we researchers need to continuously remind ourselves that we are investigating the animal *as an animal* whose way of being is existence. This is no formula; rather, it is a reminder of the task itself. (3) How can *Dasein*'s various forms be demarcated? A good starting point, I believe, would be to look across life's major evolutionary transitions that I previously noted in

Section A of Chapter 3, for example, to consider animals who are aquatic versus those who are terrestrial, animals who fly versus those who live on the ground and within the ground, and animals who are very large versus those who are very small.

In the remainder of this chapter, I turn to deuterostomes, devoting separate sections to fish, birds, and mammals. I do not discuss some fascinating deuterostomes (e.g., echinoderms, cephalochordates, urochordates) whose sublineages diverged earlier—or simultaneously with—the fish + bird + mammal lineage.

Section C. Fish

Like arthropods and cephalopods, fish are seldom considered by humans as the animals they are (see Balcombe 2016c). More often, fish are seen as food, commodities for the pet trade, entertainment for the sporting industry, or just plain bizarre creatures (e.g., some are capable of environmentally contingent sex change). Like arthropods and cephalopods, fish have been "othered" by their different-from-human looks—prominently including a seemingly expressionless face—and their near-exclusive aquatic habitat, among their many ontical differences. Even so, fish do display complex behaviors that warrant talking about fish *Dasein*, or so this section aims to show. I also note that the number of fish species (> 33,000) is approximately the same as that of all other vertebrates combined, but that extant fish considered together do not constitute a single—or monophyletic—group, unlike fish within the major subgroups (e.g., jawless fish [lampreys], cartilaginous fish [e.g., sharks], ray-finned fish [e.g., trout], and lobe-finned fish [e.g., coelacanths][see Scales 2018]). The vast majority of "fish" species today are ray-finned fish, with teleosts being its largest subgroup. Teleost fish are the primary actors in my discussion, and I refer to them simply as "fish."

Like previous sections, this one on fish is organized around *Dasein*'s way of being (i.e., existence) as seen through the lens of temporality (i.e., beenness, presencing, projection) + spatiality. Let us begin with an understanding of beenness as it follows from an interpretation of ontical biases, starting with fear. First, let's establish what it looks like to a biologist for an individual fish to be afraid (Balcombe 2016c). At the physiological level, the fish may breathe faster or release an alarm pheromone. At the behavioral level, the fish may flee, become stationary ("freeze"), try to look bigger, change color, stop feeding, or avoid areas of previous danger. Given similar fear responses in other vertebrate individuals, we should be confident that fish too can be afraid. But are they aware of this fear? Here is where memory is important because it allows the biologist to temporally dissociate the fearful event from the fear itself. Carp and pike who have been caught ("hooked") only once by human fishers have been observed to avoid bait for up to three years thereafter. And—in a laboratory-controlled situation—the crimson-spotted rainbowfish can learn to overcome an initially fearful experience—in particular, overcoming the fear of a gradually enclosing net by locating a small escape portal—and, then, without further reinforcement, to remember this mode of escape eleven months later—a period which is approximately one-third of a rainbowfish's life span (Brown 2001). So, we can be confident that fish have memory, and we can feel justified in interpreting their fear in terms of fish *Dasein*'s attunement. Of course, reductive mechanisms can always be invoked to explain the phenomena without invoking attunement, but at some point, we should particularly see "moving-goalpost" explanations as driven by anthropocentrism.

Many fish are found in proximity to other fish, and we might wonder whether they are capable of recognizing the fearful attunement of others and being empathic. Like fear, empathy can be considered an attunement, but now within an explicitly social context. But how might this occur in fish? One hypothesis—called "the behavioral copying mechanism"—states that physiological and behavioral correlates of fear observed in other fish directly "trigger" a similar fear response in the observer fish. A second hypothesis—called "fear contagion"—states (in more phenomenological terms than mechanistic ones) that the observer fish is actually aware of the other fish's fear as shared. Importantly, only the second hypothesis supports the idea of empathy as an attunement in fish.

To test these two hypotheses, Akinrinade et al. (2023) studied zebrafish under laboratory conditions. In the first phase of the experiment, the authors showed two prerecorded videos of the same demonstrator fish to the observer fish. In one video, the demonstrator fish displayed erratic and freezing behaviors indicative of being stressed. In the other video, the demonstrator displayed the normal swimming behavior of unstressed (or "relaxed") fish. Observer fish preferentially directed their attention to the video showing stressed fish. Observer fish also manifested freezing behavior. In the later "test phase," the same observer fish were simultaneously shown two videos of relaxed fish, but only one of the videos showed the same fish who had previously been stressed (in the first phase of the experiment). Now, the observer fish approached and preferred being near to previously stressed fish, even though both videos in this test phase showed unstressed demonstrator fish only. In other words, observer fish are discriminating based on remembering whether they had observed particular demonstrator fish as stressed, countering the idea that the behavior of the observers was merely an immediate response to someone else's behavior (i.e., arguing against the behavioral copying hypothesis). The observer fish's "consoling" behavior was thus interpreted as empathic (DeAngelis and Hofmann 2023), now to be considered an attunement of fish *Dasein*. Although not discussed, I note that rage may be another fish attunement (see Balcombe 2016c).

Lastly, I want to add to Murphy(2017)'s ontical claim that confining zebrafish (among other fish) to featureless (i.e., environmentally depauperate) aquaria can lead to their (medically defined) depression. At the ontological level, what we may have here is what Heidegger called "contemporary boredom" (see Table 1.1), a fundamental attunement that—in this case—has come about through humankind and technology's imposition on fish *Dasein*.

Now, let us turn to presencing in fish. Key interrelated subconcepts that foster interpretation of a fish's behavior as presencing are (1) awareness of one's world, (2) awareness of self—one manifestation of which is awareness of one's own pain, and (3) learning (associative, generalized, and social). All three point to a fish as being-amidst-entities, where "entities" point to other *Dasein*—but not to be radically distinguished from one's own *Dasein* or from non-*Dasein*. A major barrier for many people to acknowledge awareness in fish—say, relative to terrestrial mammals—results from a lack of understanding as to how sensory perception occurs in an aquatic environment. As a result, many people are either unaware or underestimate a fish's sensory capacities. In murky or dark or otherwise visually obscured aquatic environments, fish utilize two important senses—among others—to detect and localize others, senses that are largely missing from humans. One sense manifests as a fish's ability to interpret changes in the pressure of surrounding water as the distant presence of others. Embodied in the so-called lateral line of fish (and aquatic amphibians), this sense that interprets pressure changes is effective for up to a body length or more. As for human eyes, the lateral line is able to interpret a moving object as moving in one direction or another even while both the fish and object are moving, so this sense is not static; rather, the lateral line enables a dynamic interpretation that can generate further action, e.g., escaping from a nearby larger predatory fish or capturing nearby prey. The lateral line enables fish to live in environments that are largely incomprehensible to human life.

A second sense—one particularly well suited to life in the highly conductive aquatic medium but again foreign to human sensory perception—manifests as the ability of some fish (e.g., sturgeons) to interpret the weak electric fields generated by all living organisms as the presence and location of others who are nearby. Called "electroreception," this sense is closely related to—indeed, can overlap—an enhanced electrogenic ability found in some fish (e.g., knifefish or "electric eels") who actively produce stronger electric fields for long-range electrolocation and electrocommunications, and for defense against predators (Yong 2022). Electroreception enables sharks, for example, to detect fish who are hidden (i.e., invisible to eyes) in sand. Aided by their dual electroreceptive and electrogenic abilities, knifefish can be trained to distinguish between a clay pot with a glass rod and one without, between solutions of differing conductivities, and between crosses and spheres. The Ubangi elephantnose fish uses its electroreceptive abilities for localizing nearby objects and its eyes for localizing more distant, large, moving objects—like predators. Electrogenic signals are also used to claim territory and to resolve fights, analogous to how other animals use warning coloration (e.g., in octopuses) and vocalizations (e.g., in humans).

While we humans can transit many kinds of environments (e.g., the deep sea, caves, high mountains) and use our technology to detect objects in novel ways (e.g., microscopes to see minuscule creatures), we continue to dwell in *Dasein* with a limited repertoire of senses and sensory perceptions.[14] Unfortunately, in *Being and Time* and *FCM*, Heidegger dismisses the biological study of animal (including human) senses as misguided reductionism, although he nevertheless performs numerous ontological interpretations of phenomena described by biologists that are associated with human touch and sight. In making his interpretations, Heidegger admits to giving priority to the ontological; that is, the ontical is largely "made" to fit his prior understanding of the ontological. In a countervailing sense, my approach might be thought as prioritizing the ontical, particularly as understood by biologists. When the topic is animals and their ways of being, rather than being itself, there is no facile solution to this dichotomous conundrum, and despite decades of trying, Heidegger never found an adequate solution. My own approach is less ambitious: (1) acknowledge the ontical as providing all that we can have direct access to even though being is what we're trying to understand, (2) acknowledge that not all we can know must be sensory based, and (2) avoid anthropocentric thinking.

Returning to fish, I note that much more could be said about their senses than I undertake, for example, as concerns their functional camera-like eyes and their ears (the latter are located internally). But now, I turn to the matter of pain in fish. Feeling pain requires at least a minimal sense that the feeler of pain is a self, so the question of pain in fish is ontologically significant. Apparent expressions of pain by fish are often ignored, for example, whole-body flipping-and-flopping and gill opening-and-closing after a human fisher removes a fish from water (Balcombe 2016c). But even when fish do (audibly) cry out in pain, the human typically cannot hear this cry except with the aid of specialized equipment. But how do we know that these fish *really* are in pain? Here, "really"—for fish and humans—points to a criterion of reasonableness rather than some absolute determination.

So, let us consider experiments that provide reasonable tests of the hypothesis that fish are aware of their own pain. Early studies (see Elwood 2021) revealed that the injection of irritants[15] (i.e., acetic acid, bee venom) into the lips of trout (*Oncorhynchus mykiss*) results in otherwise unusual behaviors—the trout rub their lips against gravel or the side of the tank, their rate of ventilation increases, and they perform whole-body rocking movements—all behaviors reasonably suggesting a fish in pain. Additional support for the pain hypothesis comes from the observation that injecting morphine—an analgesic in humans—into the trout reduces behavioral differences between irritant-injected and mock-injected trout.

A similar study took advantage of a trout's natural propensity to avoid novel objects and to increase the rate of respiration in their presence—both interpreted as indicators that the fish is fearful (Sneddon et al. 2003). But when the lips of the fearful trout were additionally injected with an irritant, the trout no longer avoided novel objects as much, although the rate of respiration remained unaltered. This has been interpreted to indicate that the irritant-induced pain overwhelmed the fish's natural neophobic behavior. Injecting the fearful, irritant-injected trout with morphine restored normal neophobic behavior because (according to this interpretation) the trout are once again paying *attention* to their fears. Regardless of the interpretation, the complexity of these behaviors argues against the idea that simple nociceptive reflexes provide a complete explanation. Instead, these trout likely have an awareness of their environment and they integrate that awareness with an awareness of their very own pain. Related studies have been performed in goldfish (*Carassius auratus*), and there the authors conclude that goldfish weigh their (positive) consumption of food against the (negative) avoidance of electric shocks (Millsopp and Laming 2008), again supporting the idea that awareness—and self-awareness—should be part of understanding a goldfish as in pain.

Fish also manifest self-awareness in contexts other than pain. As I explained for bees in Chapter 4, documenting self-directed behavior in unusual, or completely foreign, contexts is thought to provide evidence that such behavior reflects self-awareness rather than mere reflex. Perhaps the classic way to assess self-awareness is the mirror test. The downside of the mirror test when applied to our fellow animals is that there are numerous reasons other than not being self-aware as to why it could fail, such as, lack of concern about the mark, hesitancy as concerns looking directly at another, and prioritization of some other sense (e.g., the olfactory) than the visual in recognition (de Waal 2019a).[16] Additionally, fish do not have dexterous hands for touching. Applied to fish, the mirror test would clearly have to be modified.

The most convincing case to date for mirror self-recognition in fish comes from observations of bluestreak cleaner wrasse (*Labroides dimidiatus*)(Kohda et al. 2019, 2022, 2023). Indeed, cleaner wrasse's broader behavior can be easily interpreted in the framework of all aspects of *Dasein*—beenness, presencing, projection, and spatiality—but here I focus on the self- and social-recognitive behaviors that are particularly relevant to their presencing. First, a quick summary of cleaner wrasse behavior in natural settings (see Brown 2015; Balcombe 2016c; Scales 2018): Cleaner wrasse consume mucus, scales, and ectoparasites found on numerous larger fish—called "clients"—who are not cleaner wrasse but who are individually recognized and remembered by the cleaner wrasse. Their service to their clients takes place at localized "cleaning stations," where cleaner wrasse and clients present themselves to each other via specific behaviors (e.g., a dance-like motion in the cleaner wrasse). Clients—some of which are predatory fish—tolerate, but only to a degree, the eating of their mucous and scales but they insist that ectoparasites also be eaten. If the cleaning is inadequate (e.g., too much mucus is eaten), the client leaves and does not return for future cleanings. Nevertheless, cleaning is a valued service. For example, clients tolerate cleaning in or near their own mouth more than the opportunity to make the cleaner wrasse a one-off meal. Cleaner wrasse are selective too. For example, they sometimes prefer a mucus-only meal, but this approach occurs less often with predatory clients who pose increased danger. Likewise, if there is a queue of waiting clients, cleaner wrasse may prioritize newly recognized fish over the locals, who remain in the area regardless. Cleaner wrasse also remember when individual clients last attended the cleaning station and prioritize those whose previous time of cleaning was more distant. In all this, recognition of others as the complex actors they are plays a central role for cleaner wrasse.

But what about the suitably modified mirror test for cleaner wrasse? In brief, the test was performed in a laboratory aquarium where a mirror could be conveniently positioned against a glass wall. The mark on the cleaner wrasse was made by injecting a pigmented gel—or unpigmented gel for control fish—at positions on the throat and face that could not be self-visualized. In the presence of the mirror and mark, multiple idiosyncratic behaviors were repeatedly performed. These behaviors were absent with mirror or mark alone, or with the mirror and unpigmented mark. For example, the experimental cleaner wrasse often directed the mark's position toward the mirror, followed immediately by self-scraping against gravel or rock. This scraping was directed to the site of the mark and was interpreted as an effort to remove the mark because it is recognized as foreign (i.e., not self). We might wonder how the cleaner wrasse could know that the mark is indeed foreign if the fish had never before seen her own throat. One possibility is that the cleaner wrasse recognizes the mark in the mirror as an ectoparasite, about which she is already familiar as non-self (e.g., based on her own cleaning activities of client fish), and commences the scraping behavior to rid itself (and not the mirror!) of the parasite. Supporting this hypothesis is the observation that the color of the mark must be similar to that of observed ectoparasites (i.e., brown, and not red or green) or there will be no scraping behavior. Such parasite-induced, self-directed behavior is characteristic of many fish in natural settings. These results demonstrate that the mirror test must be carried out in natural-like contexts.

As with many important research claims, the above interpretation has been challenged (e.g., see de Waal 2019a), and this has fruitfully led to further hypothesis testing. In one set of experiments, the mirror was replaced by (motionless) photographic images—either of another, unfamiliar fish or of herself. Mirror-naive fish attacked both images, but if the cleaner wrasse had already passed the mirror test, then she only attacked the image of the other fish, and not of herself. This motionless-image test argues that the experimental cleaner wrasse's behavior results in forming a memory of herself that can be independent of observing her own movement (as occurs in the mirror test) or proprioception (as occurs in a natural setting). Further experiments showed that this self-recognition was specifically directed to facial characteristics in the image.

Given all this, we can unhesitatingly say that cleaner wrasse pass the mirror test. Of course, this is not to say that cleaner wrasses' self-recognition is indistinguishable from that found in other species who also pass the test. Instead, such diversity could lead us to a deeper understanding of *Dasein*'s various forms. But such understanding remains out of reach as long as ontical diversity is understood exclusively mechanistically.

In concluding my consideration of presencing, let me very briefly describe some examples—without going into many experimental details—that document a fish's ability to learn. **(1)** Introduced by humans into the Tarn River (southwestern France) in 1983, predatory catfish (*Silurus glanis*)—like some marine mammals, e.g., orcas and bottlenose dolphins—have recently learned to "beach" themselves on the shore in order to capture nearby prey, for example, pigeons in the case of this catfish (Cucherousset et al. 2012). This behavior has not been observed in the catfish's native range. **(2)** In an aquarium setting, Atlantic cod (*Gadus morhua*) have invented and remembered two ways to release food from a feeder by pulling on the loop of a string attached to the feeder (see Balcombe 2016a). One way is to use their mouth to pull the loop. The other is to hook the loop to a tag attached to them by the experimenter (for purposes of individual identification) and swim away from the feeder. **(3)** Under controlled conditions, carp (*Cyprinus carpio*) are able to discriminate various sorts of music, e.g., jazz from classical; similar music of the same timbre; and melodies with mirror-image orders of pitch values, even (in one case), with no cues from timbre (Chase 2001). **(4)** Fish, like other vertebrates, have been shown to discriminate quantity in a manner that is not dependent on specific content but rather on the content's intrinsic numerosity (see Messina 2021). This occurs whether the fish are presented with differing numbers of, for example, food items (i.e., a spontaneous choice test) or by association between differing numbers of symbolic items attached to a subsequent food reward (i.e., operant training). **(5)** Damselfish (e.g., *Stegastes diencaeus*) farm their own algal gardens (see McGhee 2021) and, through this farming, have formed a mutually beneficial relationship with mysid shrimp (*Mysidium integrum*)(Brooker et al. 2020).

Now, some examples of social learning in fish: **(6)** cichlid fish (*Astatotilapia burtoni*) learn hierarchical relationships by watching fights between rival males, and they can then infer implied relationships (e.g., male *A* dominates male *B*, male *B* dominates male *C*, ∴ male *A* will dominate male *C*) (Grosenick, Clement, and Fernald 2007). In formal logic, these fish perform "transitive inference." **(7)** In Chapter 3, I interpreted a bee's play as a manifestation of beenness, but play must be learned and is oftentimes performed in a social context. Balcombe (2016c) relates a humorous account of a fish-and-cat game involving a lone banded-cichlid (*Heros severus*) and several house cats who started drinking from the cichlid's aquarium, making this an example of interspecies social play that satisfies Burghardt's (2005) five criteria of playful activities (listed in my Chapter 4). See also Chapter 13 of Burghardt 2005 for numerous examples of play in a wide diversity of fish. **(8)** Cooperative hunting between two species has been documented between a grouper (*Plectropomus pessuliferus*) or coral trout (*Plectropomus leopardus*) and giant moray eels (*Gymnothorax javanicus*), and between humphead wrasses (*Chelinus undulatus*) and octopuses (Octopus *cyanea*)(Vail, Manica, and Bshary 2013). Furthermore, fish prefer more effective hunting partners over less effective ones (Vail, Manica, and Bshary 2014). **(9)** When deciding whether to cheat or not, female cleaner wrasse adjust their behavior depending on whether or not their male partners—who will punish their cheating behavior—can observe them while they (i.e., the females) clean their clients; they cheat less often when they can be observed (McAuliffe et al. 2021). **(10)** When positioned above her nest, a female brown trout (*Salmo trutta*) can induce a nearby amorous male to release his sperm by vigorous body quivering, which is her typical orgasmic behavior. However, in some cases, she chooses not to release her own eggs, so her orgasmic behavior becomes a form of social deception, one that may indicate some dissatisfaction with the male (see Balcombe 2016b). **(11)** Nine-spined sticklebacks (*Pungitius pungitius*) are more likely to switch their foraging location when they observe other sticklebacks who return with more food—and not less—than they did (Kendal et al. 2009). Thus, they judge the quality of their own foraging location by observing other sticklebacks rather than by directly comparing foraging locations. **(12)** Archerfish (*Toxotes chatareus*) can learn to discriminate among computer-generated images of forty-four human faces independent of color, brightness, and head-shape (Newport et al. 2016). Relatedly, female guppies (*Poecilia reticulata*) can learn to recognize—and then to remember—conspecifics within their own schools, even though these guppies are cryptic and uniform in appearance (Griffiths and Magurran 1997).

Now let us turn to projection in fish as evidenced through an interpretation of their goal directedness. One good indicator of goal directedness is tool use, which has been documented in the family of wrasses, including the orange-dotted tuskfish, *Choerodon anchorago* (see Balcombe 2016a). This tuskfish can expose a clam hidden under the seabed's surface by directing a jet of water toward it. Then, using its mouth the tuskfish picks up the clam and carries it to a distant rock or other hard surface. Still holding the clam, the tuskfish then smashes it against the rock using a robust head flick coupled with an appropriately timed release of the clam. After repeating this smashing behavior enough times, the clam's shell breaks open, and the tuskfish has a meal, thus demonstrating persistent tool-use behavior. Additionally, the overall behavior implies planning and anticipating a future event. This is because the abilities to direct a jet of water and to carry the clam to the rock would yield little by themselves; rather, they are placed in a temporally and spatially contextualized sequence of behaviors that terminate with immediate tool-use behavior (i.e., smashing the clam against a rock) and ready access to a meal.

Archerfish also generate a jet of water which they can target so as to knock an unsuspecting flying insect down to the water's surface, thereby allowing its capture—and consumption (see Balcombe 2016a; Schuster 2018). Within the context of tool use, the water itself becomes the tool. Novice archerfish gradually learn how to use this tool by carefully observing conspecifics who are more skilled. But archerfish behavior is flexible. They also hunt in the traditional way of predatory fish, or—if flying insects are sufficiently close to the water's surface—archerfish can capture them by literally jumping out of the water and into the air.

In the previous part on presencing in fish, I mentioned cooperative hunting behavior between different species (e.g., grouper and moray eel) as illustrating social learning. Now, I want to add to this account by saying that this contingent behavior reasonably entails planning and anticipating a future event and, thus, can also be interpreted as the ontical manifestation of projection. Do we know that this is the case? Well, here we are confronted with an insufficiency of ethological studies—and particularly of *experimental* ethological studies. This general situation oftentimes provides fertile ground for introjecting an anthropocentric bias and, thereby, short-circuiting the ontical and ontological thinking that is needed. To remedy this, of course more ethological studies are necessary, but more "data" alone cannot bypass the coequal necessity of ontological interpretation.

In concluding this section, I ponder how ethological studies of a fish's (ontical) sites and activities of habitation can be interpreted in terms of fish *Dasein*'s spatiality (see *Being and Time*) or "regions"[17] of dwelling/being-at-home and non-dwelling (see *Contributions*). Taking up a favorite late-Heideggerian theme, I begin with technological thinking. Technological thinking "is" historical being as it relates to modern humans—Heidegger says this much (see *Contributions*).

But, the current epoch of historical being also relates to many of our fellow animals, although knowing this requires that we humans be sufficiently open to their ways—some different than ours—of apprehending historical being. In most cases, apprehending is more than mere thinking, except perhaps for some rarefied philosophers. Apprehending also entails practical action; indeed, we should draw no strong distinction between thinking and action. It is not hard to see ontical manifestations of technological thinking by humans in environmentally disturbed fish habitats, but more central to my concern, in disturbed fish themselves. Can fish and humans dwell (as opposed to non-dwelling) in disturbed habitats? This question must first be properly understood. In particular, there can be no necessity to dwelling or non-dwelling—"necessity" is the language of science and logic. Rather than imposing necessity, a disturbance might open fish and humans to there being possibilities, for example, to (continue to) dwell, or not. The veined octopus (see Section B) continues to dwell in new ways amidst human-disposed trash (e.g., coconut shells and broken bottles), but Atlantic cod on the coast of North America may now reside in undwelling and may continue in this way even absent further human overfishing (see Lilly 2008).

Outside disturbed habitats, dwelling may seem to be a natural state for fish, but dwelling is not something strictly natural nor does it map to a fish's typical behavior. Rather, fish—like humans—always have opportunities to engage their very being. This is why ethological demonstrations of behavioral flexibility and intelligence are important to establish. But oftentimes overlooked is the fact that flexible and intelligent behaviors—previously discussed in the context of presencing—always occur in places where a fish might dwell—where a fish might see opportunities that illuminate her very being. Reinforcing this, Heidegger says, "Seeing and hearing are distance-senses not because they are far-reaching, but because it is in them that Dasein as deseverant[18] mainly dwells" (*Being and Time*: 141).

Now, let us add another layer of interpretation to examples previously discussed in the frame of temporality's attunement. As attuned, the rainbow fish is fearful of the nearness of a gradually enclosing net (Brown 2001). However, it is not just the enclosing net that is fearful but the entire space in which she finds herself. Even before the experiment begins, the researchers know that a "successful" experiment requires that the fish accommodate to her surroundings in the aquarium. What does this mean? It means that the fish must be at-home, even if this translates to being not-at-home in the presence of a net. The alternative—the complete absence of any form of at-homeness (i.e., contemporary boredom)—may be found in the depressed zebrafish located in the featureless aquarium (see Murphy 2017). We see that the attunement of rainbow-fish-*Dasein* is inextricable from her spatiality and dwelling.

Relatedly, and now considering temporality's presencing in fish, we can also reasonably claim that their sensory perception (see my discussion of the lateral line and electroreception) and social recognition (see my discussion of this in cleaner wrasse) are not comprehensible absent there being nearness, farness, and direction toward some region of dwelling. To be clear, spatiality and dwelling are not manifest ontically simply as a background condition in which a fish happens to find herself. Rather, they point to her world—a world that includes herself—as having a particular sort of meaning. It is that world which enables her sensory perception and social recognition, and not the reverse.

Let us next consider the predatory catfish who leaps onto land to capture an unsuspecting pigeon (Cucherousset et al. 2012). In learning to do this, the catfish has now expanded her region of dwelling, even if only momentarily. We could say this is a manifestation of a "decision" about catfish being. But that region of dwelling is not simply a physical location. If the catfish were placed in the very same spot by a human fisher, then there would be no dwelling. Similar for the damselfish who farms her algal garden (see McGhee 2021). This garden is not simply a space but is also a place of meaningful relation within her world.

Lastly, a brief comment on spatiality and dwelling as they relate to temporality's moments, or aspects. Previously, projection in fish *Dasein* was interpreted through goal-directed behavior and tool use (e.g., in tuskfish and archerfish; see Balcombe 2016a), and through cooperative hunting (e.g., between groupers and moray eels; see Vail, Manica, and Bshary 2013). However, none of these interpretations are complete absent the co-primordial interpretations of their beenness, presencing, and spatiality. As concerns spatiality, tools are not tools unless they are nearby—at least initially—where "nearby" denotes not an absolute distance but a bringing-close (*Being and Time*: 142). But this is still not enough. Tool use requires a directionality—again, not necessarily in a scientific sense but in the pointing to a project's goal, which must be seen through *Dasein*'s existence and not as a material end. As for cooperative hunting, we see *Dasein*'s being-with-others in a most intimate manner, where now a goal is shared between multiple instances of *Dasein*, and where sharing also implies a bringing-close and a directionality.

Section D. Birds

Even though the number of individual birds in North America has plummeted over the past fifty years (Rosenberg et al. 2019), birds remain iconic animals in the Western popular imagination. Consider the nearly two-million American members in the National Audubon Society (https://www.theclimategroup.org/national-audubon-society, January 11, 2024) and the existence of additional American organizations specifically dedicated to the conservation of birds (e.g., Boreal Songbird Initiative, American Bird Conservancy, Partners in Flight, Ducks Unlimited, The Peregrine Fund). Ethologists, likewise, have directed considerable attention to behavioral studies of birds, and it is a tiny fraction of these to which I turn my attention in this section (see Ackerman 2016, 2020 for good nontechnical overviews of bird behavior). But as a preliminary observation, I note that dismissive comments about small bird-brain size are being countered by recent neurobiological findings. For example, the neuronal densities in brains of parrots and songbirds are approximately twice that in similarly sized primate brains, and parrots (e.g., grey parrots, macaws, keas) and corvids (e.g., crows, ravens, jays) have a higher proportion of neurons in their pallial telencephalon (which corresponds to the cortical forebrain in humans) than primates (and other birds, see Olkowicz et al. 2016). While not necessary for understanding bird *Dasein*, such brain-centric findings as these have provided justification for reductive biologists to overcome their own widespread anthropocentrism and to claim publicly that birds too can be cognitively advanced. In my view, advanced cognition (e.g., generalized learning versus associative learning) is not necessary to be in *Dasein* (see O'Donoghue et al. 2022; Sloutsky and Turner 2023; Wasserman, Kain, and O'Donoghue 2023), but for Heidegger it is implicitly built into (human) *Dasein* and is not simply correlative.

The structure of this section—with its focus on aspects of temporality and spatiality as interpreted through the ethological record of birds—is similar to that of others in this chapter. Let us begin with beenness and its relation to ontical attunements. Oftentimes confused with psychological moods and biases of an autonomous individual, attunement instead refers to *Dasein*'s preunderstanding of an entire situation when embedded in a meaningful world. As one example, the ecologist Bernd Heinrich (2019, 2024) relates his extended encounter with a particular barred owl (named "Owl") at his remote cabin in Maine. Heinrich fed Owl raw meat, and Owl quickly became a regular partaker. However, one day Heinrich had no fresh meat, so he gave Owl all that he had, which was a (dead) rancid squirrel. Owl consumed this offering, but thereafter Owl refused all food from Heinrich, even though Owl remained in the close vicinity of Heinrich's cabin and continued to display no fear of Heinrich himself. What had changed was that Owl began to ignore Heinrich; for example, he looked away in Heinrich's presence. Heinrich interpreted Owl's behavior as disdain or loss of trust. In my interpretation, Owl has called up a different attunement toward Heinrich and his food offerings. Is this thinking anthropocentric? Heinrich says, "... if we are convinced a priori that anything that looks like human behavior in another animal can't be motivated by the same or similar emotions, then how can we ever bridge the divide between us?" Of course, Heidegger would say that the divide—ontological in his case—is indeed unbridgeable, but as an a priori claim, it is fundamentally anthropocentric.

Extensive ethological observations of New Caledonian crows (discussed later under "Projection") and grey parrots (discussed later under "Presencing") can be used to interpret all aspects of *Dasein*'s temporality, but here I briefly discuss their (ontological) beenness as interpreted through their (ontical) attunements. McCoy et al. (2019) asked whether the tool-using behavior in New Caledonian crows can be its own reward or whether a positive valence is associated only with some material reward—in this case, food. To test this, the authors employed the cognitive bias test that I have previously discussed in the context of documenting emotional responses in honey bees (see Chapter 4). This test is based on quantifying responses to newly ambiguous situations, and in McCoy et al.'s study (and, now, without going into the specific methods) the authors demonstrate that tool use in the service of obtaining a food reward contributes to a positive affective state ("optimism"), and that this contribution is "above and beyond that created by simply gaining food (i.e., crows gain intrinsic enjoyment from this behavior)." In a less controlled manner but now focusing on a few African grey parrots over decades of exposure to many novel objects and novel tests, Pepperberg (2002, 2020) illustrates their incredible, long-term inquisitiveness. Here I think it likely that inquisitiveness and optimism are related descriptors of attunement.

Another attunement—one that is also centrally important in early human development (e.g., in language acquisition) and later acculturation (e.g., in fostering normative behavior), but also widespread in other animals, including many birds—comes from an interpretation of the capability to imitate others. Here, I emphasize vocal mimicking in birds. Such mimicking points toward a particular—or biased—orientation that these birds have toward their world. Reflecting a holistic understanding of *Dasein*'s world, vocal mimicking in birds also subsumes memory (= another aspect of beenness; see Zhao et al. 2019), awareness (= presencing and learning; see Audet, Couture, and Jarvis 2023), and goal directedness (= projection; see Cossins 2015). It would be much too reductive to view mimicking as just another biological capability. The lyrebird illustrates vocal mimicking most dramatically with her vocalizations of numerous other species—vocalizations that are accurate enough to fool other birds, and other animals (Ackerman 2020).

Now let us consider a more explicitly social manifestation of attunement in ravens as found in the consolation behavior displayed by bystanders of inter-raven conflict (Fraser and Bugnyar 2010). Based on documenting 152 instances of conflict, the researchers observed that within two minutes after an intense conflict, bystanders made consoling gestures to the harmed raven, such as, by sitting nearby, offering comforting sounds, preening the victim. By contrast, evidence for reconciliation between crows directly involved in conflict was weak. As in numerous other ethological studies, researchers in this study offer two possible explanations: Either the bystanders are responding only to the external signs of distress (that is, to mere behavior associated with a negative valence) or else they are displaying empathy through their recognition of a potentially shared emotional state in another raven. That involved bystanders share closer relationships to the victim than to the aggressor favors the latter explanation, since the response can no longer be to such behavior simpliciter, but this is not a definitive test according to the researchers. Additional evidence for an emphatic response has been offered for greylag geese (Wascher 2008).

Besides attunement, the other ontical linchpin in the interpretation of beenness in birds is memory. As examples, pigeons can remember homing sites used several years earlier (Collet, Sasaki, and Biro 2021). And, American crows can remember the individual human faces. This latter claim is based on observing that crows direct a scolding behavior specifically toward those humans—shown experimentally to be identified through facial recognition—who were linked to their own prior trapping (Marzluff et al 2010). What's more, this remembering can be transmitted to crows who were never trapped and who had not even observed the prior trapping (Cornell, Marzluff, and Pecoraro 2012). Consistent with this being social learning, scolding behavior in the presence of the offending humans (but not others) doubled in frequency at one site over five years of observation, even though there was no further trapping. Similar findings of valenced facial recognition have been reported in magpies (Lee et al. 2011). In a study of Antarctic skuas, facial recognition was shown to require little—and possibly no—prior exposure to humans, consistent with cognitive rather than associative learning (discussed further below; Lee et al. 2016).

A third particularly thoroughly documented example of memory is that in scrub jays (Clayton and Dickinson 1998; Clayton, Yu, and Dickinson 2001; de Kort, Dickinson, and Clayton 2005). In their studies, the authors took advantage of the fact that a natural occurring behavior, namely, caching multiple food items in multiple places and later retrieving and consuming them, also occurred under laboratory conditions, such as using cages, artificial cache sites, and researcher-provided food items. First of all, this basic behavioral observation itself demonstrates that the jays form memories of the caching locations. But what are the cognitive limits to this behavior? For example, what about caching and retrieving two different food items that spoil at different rates, like perishable mealworms and nonperishable peanuts? Do the jays retrieve the mealworms even after they have decayed? The answer is no; instead, they go for the peanuts. So, the jays remember not only where but also when they have cached particular food items. But what if jays cache the same perishable food item at multiple times? Can they remember which caches have spoiled mealworms, and which have not? The answer is yes; they only avoid caches of spoiled mealworms. So, there is an even deeper structure to their memory. But what if, mid-course in the experiment, the jays are presented with mealworms that spoil at a faster rate than those cached earlier, paralleling what the jays might find in a typical environment of variable temperature? Upon the unexpected discovery of spoiled food (based on prior experience), do the jays recalibrate their estimates of spoilage times as evidenced by avoiding other caches of the same food type? The answer is yes. So, the seemingly simple behavior by jays of caching and retrieving food is not a simple memory at all. Instead, jays have an awareness of what, where, and when—all the while remaining open to revising their behaviors based on new information.

So much for my discussion of beenness.

My next task is to illustrate that birds can presence in their *Dasein*, based on an interpretation of ethological observations demonstrating ontical awareness of environment and self, and various sorts of learning (e.g., associative, generalized, social). The ethological literature is large, reflecting a broadly based human fascination with birds, and this has pushed me to rely more heavily on generalizations found in secondary sources. For example, in *The Genius of Birds* (Ackerman 2016: 101) the author describes the complexity of bird sociality as follows:

> Many bird species are highly social. They breed in colonies, bathe in groups, roost in congregations, forage in flocks. They eavesdrop. They argue. They cheat. They deceive and manipulate. They kidnap. They divorce. They display a strong sense of fairness. They give gifts. They play keep-away and tug-of-war with twigs, strands of Spanish moss, bits of gauze. They pilfer from their neighbors. They warn their young away from strangers. They tease. They share. They cultivate

> social networks. They vie for status. They kiss to console one another. They teach their young. They blackmail their parents. They summon witnesses to the death of a peer. They may even grieve.

Ackerman then proceeds to document these statements in less general terms, but still, what is already noteworthy is how many descriptors in this quote go beyond simple scientific phraseology to an interpretation that fits with presencing in *Dasein*—"deceive," "divorce," "fairness," "tease," "console," "teach," etc. This illustrates (I believe) that—for many people—there is already an acknowledgment of such presencing in birds, and this is even more so when imagining our fellow mammals (see Section E). But particularly in the case of birds, this is not because they—unlike the arthropods, cephalopods, and fish previously discussed—are more like us. At the level of sensory perception, birds have quite different worlds which are reflected through their different ways of sensing (see Yong 2022). As examples, the field of vision for herons is 180° but that for mallards is 360°, chickens sense and process different sorts of information independently with each eye (the brain is "lateralized"), pied flycatchers have ultrafast vision, and birds in general are tetrachromats (humans are trichromats) which allow them to see what we cannot and to make color distinctions that we cannot. As concerns hearing, birds can resolve sounds more finely (faster) than we can, although this comes at the expense of sensitivity. Birds also have senses that are absent or little developed in humans. As examples, some songbirds respond in a directed manner to magnetic fields—for example, during migrations—and shorebirds called "red knots" can detect clams by interpreting pressure wave patterns (called "touch at a distance"). In sum, it seems incredible to deny that birds are aware of and immersed in a world (as Heidegger understands that concept when thinking about humans).

Now let us consider a test of self-awareness in roosters (male domestic chickens) that incorporates a procedural variant of the classic mirror test (Hillemacher et al. 2023). I have already discussed problems with applying the classic test outside humans; for example, it presumes human curiosities and is too rigidly interpreted as pass or fail (e.g., see Wittek 2021). The variant test takes advantage of a natural behavior, namely, that roosters warn conspecifics in the presence of an aerial predator by making an alarm call, but they do not do this when they are alone, indicating that they know no one (else) is then present to hear their alarm call. But do they know that they are present? The summary finding of the variant test is that—under controlled laboratory conditions—roosters make an alarm call in the presence of another rooster when a moving shadow of a hawk is projected on the ceiling, but they do not do this when they are alone (but still with the projected image of the moving hawk) or when they see their own mirror image (but still with the projected image of the moving hawk). This suggests that the rooster recognizes his image in the mirror as his own.

The proposition that birds are ontically aware of their world seems to me uncontroversial given the complexity of the environments in and through which they navigate, but still we might reasonably question whether the that, how, and what of their learning are adequate for presencing in *Dasein*. What does the ethological record suggest? I continue my exemplar approach in addressing this question and emphasize experimental rather than strictly observational studies, as experimental studies more effectively address the skeptical concerns of critics who question animal *Dasein*.

I begin with an experimental study on imprinting in newly hatched mallard ducklings to illustrate that abstract reasoning is not confined to so-called highly intelligent animals (Martinho III and Kacelnik 2016). In the training part of this study, different ducklings were imprinted on "a pair of objects that were either the same or different in shape or color." Then, in the experimental part, which used a pairwise-choice protocol, the researchers demonstrated that the ducklings who imprinted on either "same" or "different" for shape or color chose likewise when presented with the pairwise choice, and they did this even though the particular shape(s) or color(s) of the objects in the experimental part did not match that on which they had been imprinted. In other words, the ducklings had formed preferences for "same" or "different" that were independent of particular shapes and colors. The authors argue that this demonstrates abstract conceptual reasoning in a "seemingly rigid and very rapid form of learning." Of course, this is not to claim that reasoning in ducklings is mechanistically identical (or as complex, or …) to that, for example, in humans. But we should also be careful not to make humans the sole standard for inter-species comparisons, as though that encompasses all of the important questions about our fellow animals. Such an anthropocentric approach tends to preclude thinking about what it might mean for *Dasein* to be in other animals. Heidegger is clear that the biological human (organism) is not *Dasein*; rather, the human in *Dasein* is an interpretation of human behavior in thinking and acting. We need to be open to making a similar interpretation from the behavior of our fellow animals.

In another experimental study, Bugnyar, Reber, and Buckner (2016) "suggest that ravens can generalize from their own perceptual experience to infer the possibility of being seen" by observer ravens. An ability to generalize is related to—but not identical with—Heidegger's usage of "as" when talking about *Dasein* and world (see discussion of *FCM*, §69, in my Chapter 1, Section D), so this study of ravens is of interest to my project on animal *Dasein*. The actual experiment incorporates into its protocol a raven's natural behavior to cache food—including under the authors' laboratory conditions—and to do this more surreptitiously when being observed by another raven. This is because ravens steal food from each other's food caches. Among ravens in the wild, "surreptitious" means that they cache their food more quickly, take advantage of visual barriers to observing their caching, and may delay caching or delay returning to their cache until the observers are gone. The particular question addressed was what a test raven would do when there was a means for an observer raven to see the test raven's food but there was no means for the test raven to see the observer. (In fact, in this experiment there was no observer, only a recording of a raven was played.) Must the test raven see the other raven in order to act surreptitiously (i.e., is a raven merely a behavior-reader) or is the very *possibility* of being observed sufficient (i.e., can a raven generalize from his/her own perceptual experience, and possibly even have a Theory of Mind)? The results suggest the latter.

Researchers now recognize that numerous birds of the taxonomic family Corvidae, including the ravens just discussed, are particularly intelligent. Another corvid—namely, the highly social Australian magpie—demonstrated both cognitive skills (and possibly flexible ones) and prosocial skills (i.e., cooperation, and possibly altruism) in a pilot field study designed to assess the effect on behavior of attaching a tracking device (Crampton, Frère, and Potvin 2022). What the researchers observed was that magpies without their own tracking devices successfully removed the devices attached to other magpies within minutes to hours of the initial attachment. As an aside, magpies also engage in play (e.g., see Chapter 7 in Ackerman 2020; also, search "magpie play" on web browser for striking videos). Such play I have interpreted as a manifestation of beenness (see my discussion of attunement-as-play in bees in Chapter 4). Shortly, I will feature corvids in the context of projection.

I conclude my discussion of presencing by turning to another group of particularly intelligent birds—the parrots (Psittaciformes; see Safina 2020 for a general and accessible introduction to their behaviors). Keas (members of the parrot family) have been shown to reason in the presence of uncertainty by using true statistical inference (i.e., by estimating relative frequencies of reward-conferring and -unconferring tokens) rather than by estimating absolute quantities, something that had been demonstrated previously only for humans and other great apes (Bastos and Taylor 2020a,b). Furthermore, the keas could correctly revise their inferences when the experimental apparatus was subtly modified, and they could even evaluate the human facilitators' biases in the experiment and adjust their own inferences accordingly. Taken together, these experiments argue for domain-general (versus domain-specific) learning in keas.

Viewed through an evolutionary lens, intelligence in particular birds—or in any other animal, including humans—always takes place within a particular context. Analogous to what Heidegger writes about in *Being and Time*, this context is an ontical world of relationship to others—a context that is both ecological (i.e., temporally contemporary) and evolutionary (i.e., temporally past). Typically, evolutionary contextualization happens by making comparative inferences across modern-day individuals of different species, but in so doing, the study of evolution returns us to the present-day lives of individuals. Here is one experiment-based example that fits within the broader context of the evolution of intelligence (Chen et al. 2019): A female budgerigar (also of the parrot family) changes her mate preference after observing another male who is

> able to solve extractive foraging problems. This preference shift did not occur in control experiments in which females observed males with free access to food or in which females observed female demonstrators solving these [same] extractive foraging problems.

Here, we see potential positive feedback between better problem-solving skills and enhanced evolutionary fitness. Although the problems faced by modern-day humans are different, we should not be averse to testing this particular context within our own lineage.

Particular behaviors in parrots can be learned by observing other parrots perform a novel (to the observer parrot) action; that is, learning can be social. In one study, sulfur-crested cockatoos (also of the parrot family)—now common in the suburbs of Sydney (Australia)—have learned to open the lids of trash containers in order to scavenge for food (Klump, Martin et al. 2021). What is particularly interesting about this complex, five-step behavior is that it was first observed at only three suburbs, but within a few years had spread to forty-four. Furthermore, different solutions to some of the five steps were observed, and these variants correlated with particular localities within the suburbs of Sydney. The overall conclusion is that a particular behavioral innovation arose in response to anthropogenic change (i.e., to the appearance of trash containers as the parrots ventured into the suburbs), that the innovation then spread across a population of cockatoos over both time and space, and that further innovations to the original innovation occurred. In a related laboratory-based study, Goffin's cockatoos were able—above chance level—to choose and insert the correct shape of key to fit into a keyhole, and to do this without reward from the experimenters (Habl and Auersperg 2017). When dealing with more complex shapes of keys, the cockatoos would even rotate the keys in their beaks to facilitate insertion of the key. Success at this task requires that the cockatoos use an external landmark (i.e., the key shape) for orientation, a behavior that is viewed as a cognitive step beyond orientation merely to one's own body.

Lastly, I turn to the current "poster child" of bird-intelligence studies—the African grey parrot. Given constraints of space, I point to a few interesting studies only. Schloegl et al. (2012) show that African grey parrots can use inferential reasoning. Brucks and von Bayern (2020) show that African grey parrots voluntarily and spontaneously offer food tokens to other African grey parrots with whom they are familiar, and that this behavior is reciprocated. Brucks et al. (2022) show that four parrot species (3 macaws + the African grey parrot) avoid eating an immediately available food reward if, by waiting, they then receive a higher-quality food reward, and that African grey parrots in particular tolerate the longest delay in receiving the reward. Delayed gratification indicates self-control or—stated more to the point—that there is a self to control (see also Pepperberg and Rosenberger 2022).

The most extended study of individual African grey parrots has been carried out by the biologist/psychologist Irene Pepperberg and colleagues (e.g., Pepperberg 2002, 2020). In particular, the cognitive and communicative abilities of the parrot named "Alex" were explored for thirty years before Alex died, and such intensive studies of a particular individual are being extended to others with comparable conclusions, along with new insights as novel experiments are performed. Previously, I have challenged the idea that associative learning by itself necessarily precludes there being a world for an animal in *Dasein*, but still, I acknowledge that it seems easier to imagine a world where an animal's behavior is considered within the frame of mental representation, rule-based learning, and symbolic processing.

So what about African grey parrots? In arguing against exclusive associative learning, Pepperberg (2020) says,

> The parrot, Alex, by showing that he could—even at that early stage in his training—vocally indicate different attributes of a single item (its color, shape, material, and overall label), demonstrated that, even for a novel item, he could interpret the various possible questions that could be posed (i.e., determine which attribute was being targeted), search his repertoire for the set of labels that were hierarchically organized under that attribute (e.g., if the question was "What color?," know to examine labels such as yellow, blue, etc. instead of paper, cork, wood, etc.), cho[o]se the one appropriate label, and then encode it vocally.

Challenged with experimental tasks-to-accomplish that bear little relationship to a normal parrot life, Alex (and other African grey parrots) may indeed require more training (i.e., learn more slowly) than most humans, but still, Pepperberg claims, eventual success requires that Alex have the appropriate cognitive capacity. For example, Alex was able to think abstractly about color by applying a learned color to a novel object (i.e., one of a different shape, size, material). Alex could also reason in terms of relations (e.g., bigger/smaller, lighter/darker) rather than solely in absolute terms (e.g., same/different). Like humans, Alex was susceptible to the Müller-Lyer illusion, indicating not only that Alex and humans similarly perceive the world but also that we similarly process *symbolic* representation, that is, of the arrows with their differently oriented fins in this particular illusion. Alex was also adept at symbolic number representation, including addition, the concept of zero, and the use of Arabic numerals. Furthermore,

> Alex exhibited clear—if limited—sensitivity to order in various non-numerical aspects of his vocal symbolic behavior (e.g., his ability to segment, his use of sentence frames "I want *X*" and "Wanna go *Y*," where he knew that an object "X" must follow "want" but a location "Y" must follow "wanna go"). … Another parrot, Griffin, has also demonstrated some rule-governed behavior, as shown during a study of his acquisition of label order. However, evidence for formal syntax, as opposed to simple rule-governed behavior (what can be seen as a proto-syntax) is still lacking.
>
> (Pepperberg 2020)

Pepperberg's (2020) conclusions illustrate how scientific methodology is linked to fundamental claims about animal cognition. She says,

> My overall conclusion is that differences *do* exist among various species' abilities … but that when individuals of these species are given appropriate training, the differences are not as great as we once may have thought. The training cannot, however, merely be with respect to simple associations between a limited number of labels and their corresponding items, but must be rich enough to

> encompass concepts and enable the subjects to transfer the learned concepts to novel situations.
>
> From the studies I have described, it is obvious that I am not against using human-based tasks to test non-humans. My data do, however, suggest the existence of inherent biases in such tasks, given how training non-humans on symbolic representation turns out to be so important for their ability to succeed on these tasks. As researchers, we need to be aware of such issues when making claims about non-human intelligence.

As concerns parrots, presencing, and *Dasein*, it is clearly time to move beyond Heidegger's thinking.

Let us now consider birds, projection, and *Dasein*. The ontical correlate of *Dasein*'s projection can be captured in a bird's present concern for some meaningful future. American crows may or may not be aware of their own eventual death (demise), but they are aware that the death of others within their own species (conspecifics)—but not within other similarly sized species (heterospecifics such as pigeons)—is meaningful (Swift and Marzluff 2015). In particular, individual crows respond to the sight of dead conspecifics by "mobbing" (i.e., by congregating with other crows in response to the actual or presumed presence of a potentially dangerous predator such as a hawk or nearby human, and sometimes then attacking or harassing the predator) and by increased hesitancy to approach nearby food. When directed against specific humans and their behavior, mobbing behavior can reoccur over a six-week period, even in the absence of further provocation. Of course, memory, recognition, and associational thinking are involved—all features of beenness and presencing—but we should also remain open to the crows' awareness of their own fate as prey—a feature of projection.

Projection may be most clearly manifest through a bird's multistep behavior that leads to a material construction, as such an activity oftentimes requires planning and evaluation. Bower construction by the male spotted bowerbird is one such complex activity, with the completed bower having the sole purpose of attracting a receptive female for mating (see Ackerman 2020). The bower serves as a visually appealing avenue of splendor for the female; it does not become her nest. Bowers have attracted the interest of behavioral biologists because of their elaborateness, their differences in design and decoration even within species, the opportunism reflected in the selection of building materials, their built-in illusions, and their aesthetic qualities—all features which suggest that male bowerbirds are able to project in a world that includes others, and that female bowerbirds—through their comparative bower discriminations—must also have a strong evaluative sense.

Another example of construction as projection is found in birds who use tools. Tool use is widespread in the animal kingdom, including across birds (e.g., herons, kingfishers, gulls, vultures; see Ackerman 2020). One particularly striking example of tool use is that of the fire-foraging raptors collectively called "firehawks"

(Bonta et al. 2017). These predators pick up burning embers (carefully!) and drop them in a nearby nonburned area in order to start another fire and thereby flush out small prey.

However, the most widely recognized birds who use tools are corvids, including the Hawaiian crow (Rutz et al. 2016) and the common raven (Jacobs and Osvath 2023). Beyond using tools to access an immediately available food reward, ravens have experimentally demonstrated their ability to plan for future tool use and for future food rewards, indicating that planning can be sensorily disconnected from immediately receiving a reward and that it can serve distinct behaviors, that is, accessing food and accessing tools (Kabadayi and Osvath 2017). Such cross-temporal behavioral flexibility (imagination?) should compel us to consider ravens to be in *Dasein*.

Researchers make a distinction between the use of tools and the making of tools because the latter is considered to be even more cognitively demanding. To illustrate complex tool construction (and use), I turn to studies of New Caledonian crows (hereafter: "NC crows"), who—under natural conditions—can modify forked plant stems to have a (functionally significant) single hook at the end and then to use this as a tool to extract various invertebrates from tree holes and crannies. As a comparative aside, learning how to make hooked tools was a major innovation of Middle-Stone-Age humans (cited from Klump, Sugasawa, St Clair, and Rutz 2015). NC crows construct other types of hooked tools under natural conditions, but they also find and use twigs and leaf petioles—without modification—as nonhooked tools (see Hunt 1996; Hunt and Gray 2002). Under experimental conditions, NC crows can make hooked tools with novel, investigator-provided materials (e.g., a wire; Weir, Chappell, and Kacelnik 2002).

First, let us consider the *use* of novel tools separately from their manufacture. NC crows do not ordinarily use stones as tools, but they have learned—under experimental conditions—to do this in order to raise the water level in a container on which out-of-reach food is floating (Taylor et al. 2011). Importantly, they preferred large stones over small ones; they quickly learned to drop stones into water-filled containers rather than those filled with sand or air; and some NC crows immediately (i.e., without prior experience) chose denser objects (i.e., stones and rubber blocks) over less-dense-than-water objects (i.e., polystyrene blocks). Taylor et al. (2011) conclude:

> Our findings show that New Caledonian crows do not have a narrow-banded ability to process causal information that is restricted to sticks and stick-like objects and the interactions surrounding these items. Instead, this species quickly processes causal information about novel interactions between new tool types and the environment, without the acquisition of such information affecting their understanding of established tool behaviours. The differences between the causal and arbitrary tasks that we presented to the crows strongly suggest that cognitive mechanisms other than simple associative learning are involved in this processing of causal information.

Additional studies demonstrated that NC crows are able to reason about hidden causal agents (Taylor, Miller, and Gray 2012).

Now, let us turn to the *manufacture* of novel tool types. With no cueing by the experimenters, NC crows were able to combine up to four short dowel segments (and other foreign objects) in order to construct a rod that was long enough to push a food reward out of a tube and, for some crows, to complete this untrained task within four to six minutes (von Bayern et al. 2018). Such construction of compound tools had not been reported outside hominids, for example, captive orangutans (Lethmate 1982). And, human children younger than five are unable to make these kinds of compound tools (as cited in Ackerman 2020: 113). In reflecting on their results, von Bayern et al. (2018) argue that learning in NC crows is not based on accidental discovery or trial-and-error coupled with reinforcement; instead, it reflects a "higher" cognitive process. Is there further evidence for such a claim? Similar to the ravens already discussed, NC crows demonstrate an ability to plan their use of tools, in their case, by selecting the most appropriate tool for an expected future task (Boeckle 2020). Perhaps not surprisingly, NC crows safekeep for later reuse those tools that are more complex to manufacture and that are more efficient at task completion—typically, hooked tools over nonhooked tools (Klump, St Clair, and Rutz 2021).

The advanced cognitive capabilities manifested in tool making and use by NC crows make less contestable the claim that they are in *Dasein*. Of course, their particular form of *Dasein* is quite distinct from that in humans (e.g., see Holzhaider et al. 2011; Kenward et al. 2011). Still, we should remain wary of making humans the standard, particularly since cognitive capabilities are so variable within humans. "Awareness of one's own being" is a flexible term, but at least some birds have structural elements that make this claim eminently reasonable.

Lastly, I consider spatiality. Thinking the *there* of *Dasein* is always incomplete when thought in exclusively temporal terms, as though a site of meaning could be incidental to that meaning. Here, "site of meaning" points both to the who in *Dasein* and to the who and what in *Dasein*'s world (i.e., to other *Dasein* and non-*Dasein*). Heidegger makes good sense when he explains how a site of meaning points to a particular configuration of relationship in *Dasein*'s world that he calls "dwelling in being," but unfortunately he is averse to thinking of this configuration as connected to biogeography, perhaps because he fears he would be misunderstood as a crypto-scientist. Instead, what Heidegger does do is prioritize his own abstract way of understanding spatiality.

If we want to know how embodied birds dwell in *Dasein*, we cannot avoid interpreting their natural history. A common impression that humans have of birds is that, by being able to fly, they can dwell in the sky, as well as dwell for variable periods on terra firma. At one extreme, the common swift can spend many months of uninterrupted time in the air. Some birds are even deficient in walking or hopping on land, for example, common loons and Anna's Hummingbird. Still, all birds remain connected to land, for example, as chicks or when roosting. At the other extreme, some birds are flightless and reside entirely on land, for example, the emu. Still other flightless birds are largely aquatic and use their wings—now called "flippers"—to swim, for example, penguins. Other birds can both fly and swim, for example, puffins (see Masud, Mantia, and Dabnichki 2022; Smith et al. 2022). Of course, there are many other ways to characterize the 11,000 species of birds, but the means of locomotion I have chosen broadly hint at the different possible places and experiences of being an individual bird in *Dasein*. At a less broad level, each and every one of my previous examples for the three aspects of temporality in birds has already provided examples of spatiality (e.g., sites of food caching in scrub jays, sites of tool construction in NC crows, places of social interactions, places of experimental testing by human investigators, etc.). This is because temporality and spatiality form a unity in time-space.

Section E. Mammals

My fifth and final focal group is animals in the class Mammalia which, among its approximately 6500 total species, includes our own species and other great apes. (For a useful evolutionary framing of the different mammals discussed in this section, see Foley et al. 2023 and their recent phylogeny of the largest subgroup of mammals, the placentals.) While sharing a basic brain architecture (see Kaas 2009), placental mammals are morphologically and behaviorally diverse. As examples, elephants have a large trunk consisting of 40,000 individual muscles whose coordinated actions enable incredible sensory awareness and mechanical control of the environment (Dagenais et al. 2021); bats have wings and can fly and navigate in a completely dark enclosure using sonar (see Yong 2022); whales have unique ways of making sound and communicating with others, but the sounds themselves bear striking similarities to human language and music (Madsen, Siebert, and Elemans 2023; Arnon et al. 2025); and various great apes have language-like abilities that challenge the abyssal uniqueness of human language and communications (de Waal 2016a; Snowdon 2017). But beyond individual capabilities, whole-personality measures have been assessed for chimpanzees (see Altschul et al. 2018) and orcas (see Úbeda et al. 2019) for comparison with humans. These measures include: agreeableness/aggressiveness, conscientiousness (self-control, delay of gratification, thoughtful planning), extraversion (sociability, activity), neuroticism (fearfulness, vigilance, emotional reactivity), openness (curiosity, originality, tendency to find novel ideas and situations appealing), and dominance (competitive prowess, social competence, fearlessness). What we quickly intuit from reading the scientific literature is that many mammals must have diverse and rich ontical lives and, furthermore, that interpreting them (ontologically) as abyssally distinct from one particular mammal—namely, the human—is largely unhinged from any reasonable understanding of that literature. While this may seem more obvious for our most closely related mammals (e.g., the other great apes), any strong bias of this sort is unwarranted.

In *FCM*, Heidegger consciously avoided discussing our most ontically similar creatures because doing so—he thought—would confuse the reader as regards their "clear" ontological difference from humans. Of course, this is—and remains—a logical possibility, but from what we now know of the histories of various peoples and of animal evolution, we can say from a scientific perspective that Heidegger's ontical understanding of our fellow animals is limited at best, and that this leads him to make questionable ontical claims and ontological interpretations. Prominent among Heidegger's failings was his unwillingness to fully grasp von Uexküll's ecological arguments. Unlike Heidegger and some reductive biologists, von Uexküll dissociated the then current understanding of instinct as mere mechanism from his understanding that each animal has a world (*Umwelt*) in the dual sense that it is embedded in an environment and that "something" of that environment appears to the animal. As I have already shown, even the old instinct argument is largely misguided, since many "instincts" have been shown to be responsive to the external environment and/or self, leading to the conclusion that behavioral instinct is a misleading abstraction.

The organization of this section again focuses on aspects of temporality and spatiality, now understood by interpreting the mammalian ethological and neuroethological literature. Relative to other sections, here I rely more heavily on citing references rather than describing experiments in their convincing detail. This is because many people already acknowledge that other mammals share numerous important capabilities with humans, thereby making the interpretation that many, or all, mammals are in *Dasein* much more obviously compelling.

Let us begin with beenness by considering mammalian attunements. A lifetime of studies by Jaak Panksepp, who was a pioneer in the field of affective neuroscience, provides our entryway (Panksepp 2011a,b; Panksepp and Biven 2012; see summary by Davis and Montag 2019). Panksepp extended the study of emotional affects from humans to other mammals—particularly rats[19] and mice—using three approaches: (1) identifying structurally homologous subcortical Brain/Mind[20] regions, (2) documenting specific affective Mind/Brain behaviors under variable environmental conditions, and (3) experimentally stimulating or inhibiting specific affective Mind/Brain behaviors through interventions of specific Brain/Mind regions (e.g., localized electrical stimulation, pharmacological challenge, physical lesion). Panksepp considered this methodological triangulation to be a gold standard for any claim of homology of emotional affects, in particular, between our fellow animals and humans. Notably absent from his approach is the necessity for assessment of one's mental state through language, something that is commonly assumed by humans to provide evidence of that state in humans. Panksepp's approach emphasizes that the absence of full-blown human language in our fellow animals should not be equated with the absence of homologous emotional affects.

Over his career and based on many observations and experiments, Panksepp identified seven emotionally affective systems that are manifest in nonidentical forms across all mammals, indicating that they are evolutionarily ancient. He capitalized the terms for these brain-based affective systems to distinguish them from the commonly understood use of homonymous emotional words. Those systems with a generally positive valence are CARE, PLAY, and LUST. Those with a generally negative valence are FEAR, RAGE, and PANIC / GRIEF. And with either valence: SEEKING. For Panksepp, "affective systems" is a scientific phrase that refers to components of the Brain/Mind, but for my own purposes, I think it appropriate to expand his usage to considering them as attunements of *Dasein*; that is, to see Panksepp's systems as *interpretations* of *Dasein*'s behavior, and not only as components of the BRAIN/MIND. As Panksepp shows, these systems are accessible to classical operant and instrumental conditioning, to largely unconscious behavioral and emotional habits, and—in some cases—to cognitive executive thought, planning, regulation, and memory. So, while they are ancient, that does not mean that they are inflexible, for example, in not being responsive to contingent present or possible future conditions. In other words, these attunements do not determine or cause behavior.

In brief, here is what Panksepp and Biven (2012: 34–7) say about the seven affective systems:

- "The SEEKING, or expectancy, system is characterized by a persistent exploratory inquisitiveness.
- "The RAGE system … causes animals to propel their bodies toward offending objects, and they bite, scratch, and pound with their extremities. Rage is fundamentally a negative affect, but it can become a positive affect when it interacts with cognitive patterns, such as the experience of victory over one's opponents or the imposition of one's own will on others who one is able to control or subjugate.
- "The FEAR system generates a negative affective state from which all people and animals wish to escape.
- "When animals are in the throes of the LUST system, they exhibit abundant "courting" activities and eventually move toward an urgent joining of their bodies with a receptive mate, typically culminating in orgasmic delight—one of the most dramatic and positive affective experiences that life has to offer.
- "When people and animals are aroused by the CARE system, they have the impulse to envelop loves ones with gentle caresses and tender ministrations. Without this system, taking care of the young would be a burden. (See also Sato et al. 2015 for a related discussion of altruism in rats; see Burkett et al. 2016 for possible empathy in voles.)
- "When overwhelmed by the PANIC / GRIEF … system, one experiences a deep psychic wound—an internal psychological experience of pain that has no obvious physical cause. … The PANIC / GRIEF system helps to facilitate positive social bonding.
- "The PLAY system is expressed in bouncy and bounding lightness of movement, where participants often poke—or rib—each other in rapidly alternating patterns. … The PLAY system is one of the main sources of friendship." (See also Reinhold et al. 2019 AND Rygula, Pluta, and Popik 2012; see Laumer et al. 2024 for play in great apes; see de Waal 2019b for laughter in primates.)

Panksepp was particularly taken by the observations that decorticated (i.e., brain cortex surgically removed) rats still express all seven systems, and that anencephalic (i.e., born without cerebral hemispheres) human babies still display rich emotional lives. He took these observations to indicate that cognitive processing, including language, is not necessary for emotional affects, although (based on other observations) such higher-level processing could inhibit or disinhibit the affects. Similar results obtain for aphasic human stroke victims who neither speak nor think in words. Panksepp's broader point is that these affective systems participate in allowing mammals to interpret their worlds, just as we would expect attunements to do.

Attunements function in parallel with memory formation (e.g., McGaugh 2013; Ramirez et al. 2015), so now I want to focus on studies of mammalian memory. Panksepp and Biven (2012: 208) note that memories are not only fixed records of past events. In psychological terms, events which have been consolidated into primary memories can be reconsolidated in light of future events, including in rats (see Jones et al. 2012). Memories can also be of more than discrete events. For example, Kano and Hirata (2015) showed videos to chimps and bonobos while tracking their eye movements to demonstrate that, upon reshowing the same video twenty-four hours later, they actually anticipated particular exciting events, for example, the appearance of an aggressive ape-like character. This required that their memory be not only of single events but also of movie-like events. There is also evidence—for example, in dogs (Fugazza, Pogány, and Miklósi 2016)—that memories can form incidentally, that is, without the expectation of recall, an ability that is more consistent with episodic-like memory (i.e., memory of specific events in one's life) than semantic memory (i.e., memory based on learned rules). According to psychologists, the presence of episodic memory strengthens the case for an animal's self-awareness. Allen and Fortin (2013) have suggested that episodic memory is ancestral to all mammals and birds, and possibly all vertebrates. Lastly, I note that memories in various mammals can also extend over many years, even decades; for examples in bonobos and chimps, see Lewis et al. 2023 AND de Waal 2019b; for dolphins, see Bruck 2013.

Now, let us see what (a tiny portion of) the ethological record reveals about the possibility of presencing in mammals. First, a reminder. For there to be presencing, the animal must have an awareness of his/her world. Blindly moving in an environment does not qualify. Ben-Haim et al. (2021) utilized a "cross-over double dissociation" approach that—in humans—allows the experimentalist to disentangle conscious perceptual awareness from nonconscious processing, and then extended this approach to rhesus macaques (a genus of Old World monkeys). The basic idea in the approach's application is that the experimental subject behaves in opposite ways depending on whether or not he/she is consciously aware of a stimulus. The result was that the macaques performed "remarkably similar" to humans in situations where humans could also verbally indicate that they were consciously aware of the perceived stimulus, as well as in situations where humans were not consciously aware and, thus, behaved differently. This study thus provides striking experimental support for conscious visual awareness, as well as nonconscious processing, in macaques and humans, whose last common ancestor was at least twenty-five million years ago. I want to be clear here that I am not equating consciousness with awareness, only that Ben-Haim et al.'s psychological studies with macaques and humans provide evidence for their perceptual awareness of the environment and make reasonable the interpretation of that awareness within the context of macaque and human *Dasein*.

What about awareness of self? Without going into details, some great apes, the Asian elephant, and bottlenose dolphins pass the mirror test; and—with more limited implications—great apes, monkeys, and sheep (as well as some crows and wasps—already discussed) are able to recognize individual faces (see Morrison and Reiss 2018 + references therein; de Waal 2016a,b). When we see that a diversity of mammals have awareness of the environment and of themselves, we can consider it reasonable that they are in *Dasein*.

Now, let us turn to other aspects of *Dasein*'s understanding. In humans, understanding can be manifest in having beliefs about the world and sharing those beliefs with others, who in turn can share their own differing beliefs. But such sharing is largely a linguistic phenomenon, and how can one show that other mammals without full-blown human language may also have such beliefs about others' beliefs? One compelling approach—originally used to test cognitive development in (prelinguistic) human infants—is premised on the idea that the behavior of an infant—for example, the infant's direction of gaze—differs when that infant observes an adult human subject who acts on a false belief rather than on a true belief. (Here, a *true belief* of the subject is one that matches that of the infant, where both subject and infant have the same information and act rationally on it. A *false belief* of the subject is one that does not match that of the infant; in this case, the infant has more complete information than the subject.) This approach has been extended to chimps, bonobos, and orangutans as observers of adult human subjects, with modifications in the protocol that take into account the apes' greater physicality by substituting helping behavior for eye gaze (Krupenye et al. 2016; Buttelmann et al. 2017). The key result is that the three apes anticipate that the subjects have false beliefs and will help the human subject accordingly. Buttelmann et al. (2017) cautiously conclude:

> The current results provide another indication that great apes might have some level of false-belief understanding, and they are the first to demonstrate that apes are able to use this understanding in their social interactions. If supported by further research, the apparent difference between great ape and human social cognition would thus lie not in their basic capacity to "read" other minds, but elsewhere.

Another example of understanding in primates: Capuchins (a subfamily of New World monkeys), like humans, are "inequity averse" (Brosnan and de Waal 2003). This was shown experimentally by training female capuchins to exchange tokens (in this case, small rocks) with as human experimenter for food rewards, that is, either cucumbers or the more desirable grapes. When multiple capuchins received tokens in the presence of each other, the exchanges continued as usual if the reward was the same for all. But if some capuchins received cucumbers and others received grapes in the presence of each other, those who received cucumbers ceased trading and even refused to eat the offered cucumbers, which under the equal-trade condition is highly unusual. In other words, individual capuchins compare their own rewards with others and they dislike inequity.

We might wonder whether awareness of false beliefs and inequity aversion point to mammals other than ourselves being moral creatures. Frans de Waal (2006) has articulated "three levels of morality" (p. 168). At the first level ("moral sentiments") are capacities for empathy, reciprocity, fairness, and harmonious relationships. These are found across primates. At the second level ("social pressure") are capacities for cooperative living and rule following. These, too, are found in other primates, although—de Waal says—rule following is less systematic and less group oriented. At the third level ("judgment and reasoning"), de Waal says, "Others' needs and goals may be internalized to some degree, but this is where the similarities end." In thinking about these levels, we might say that moral decision-making and its antecedents play out differently across primates but that there is a substantial core of similarity that follows from their shared evolutionary history.

As the previous two paragraphs illustrate, learning in primates is oftentimes social learning, or learning from and with others. Consider orangutans and chimps, who can no longer be denied their status as culture forming based on an interpretation of their numerous, geographically variable behaviors (Vogel 1999; van Schaik et al. 2003; de Waal 2016a; Gunasekaram et al. 2024). Van Schaik et al. usefully articulate four elements that identify culture: the use of (1) labels such as food preferences, (2) signals that reflect arbitrary behavioral variants such as making particular sounds or motions, (3) skills such as particular tool use based on social transmission, and (4) symbols that become "membership badges of the social unit." Only in symbol use do humans manifest far more complexity, but even here, the difference is not absolute. Cissewski and Luncz (2021) cite evidence that all extant great apes—as well as marine mammals, dogs, and parrots—have "symbolic competencies," although these are obvious only under experimental conditions, at least for the great apes. It is important to emphasize that all four elements—labels, signals, skills, and symbols—are themselves interpretations and not simply observations, even for scientists, and that these elements can comfortably be fitted into talk of *Dasein*. But this is not to say that biology can be avoided. Some communicative gestures made by chimps and bonobos can be accurately understood by humans absent any context beyond seeing the gesture itself, strongly suggesting a common evolutionary origin (Graham and Hobaiter 2023) that precedes language features such as syntax (Crockford, Wittig, and Zuberbühler 2017).

For cultures to arise, their members must help others and be able to resolve conflicts. In great apes, including humans, conflicts inevitably arise—for example, between same-sex rivals, friends and family members who share resources, and casual in-group acquaintances who cooperate on some task—and these conflicts have to be resolved in a manner that is partially situation specific, that is, beyond some algorithmic-based response. Chimps and bonobos, but also monkeys, have developed different strategies and behaviors of conflict resolution—for example, male chimps must learn from others how to control their own considerable aggression, while female bonobos have a more direct role in teaching their males to control aggression. The important point is that these strategies are manifest socially and demonstrate an awareness of their place in a world that includes others somewhat like themselves (de Waal 1990; Chapter 10 of de Waal 2022). One naturalistic way to frame conflict resolution within a social group—as opposed to a collection of individuals working toward a common goal—is to say that the group as a group benefits when individual members sacrifice their own fitness for that of the group. (For a discussion of this as applied to chimps, see Langergraber et al. 2017.)

But do nonhuman creatures help each other? In fact, adult humpback whales have been observed to deliberately mob and harass orcas when the latter are attacking other mammals (Pitman et al. 2017). This successful anti-predator behavior continues even after the humpback whales discover that the orca's target is not a humpback whale calf but a seal, for example. The author says, "… there was no apparent benefit to humpbacks continuing to interfere when other species were being attacked. Interspecific altruism, even if unintentional, could not be ruled out." Similarly, chimps in a laboratory setting sacrifice a portion of a desirable treat in order to provide that treat to an unrelated chimp, particularly when the other chimp had previously provided assistance (Schmelz et al. 2017). And, chimps in the wild participate in risky border patrols even when it is not their own kin that they're protecting (Langergraber et al. 2017). Again, we see striking parallels with humans. In a related manner, bonobos in the wild have been observed to cooperate (e.g., in grooming and food sharing) with unrelated (i.e., of low genetic relatedness) bonobos who are in other social groupings (Samuni and Surbeck 2023). Cooperation with unrelated members of one's species for no immediate benefit is thus not unique to humans; rather, we might say that this characteristic has been amplified in the human lineage. Altruism aside, long-term interspecific relationships, sometimes known to involve particular individuals and not only their groups, have been documented between different species of dolphins (Elliser and Herzing 2016) and between gorillas and chimps (Sanz et al. 2022). Minimally, we can say that these animals are aware of others and derive some sort of meaning from their interactions. Not surprisingly, social groupings themselves are flexible. For example, Danaher-Garcia et al. (2022) have documented the partial merger of two dolphin social groups, both previously observed to be stable for at least twenty years, after one of these two groups migrated into the range of the other.

For Heidegger, projection is a prioritized aspect of being which identifies individual *Dasein* as running ahead unto death for the sake of itself. How might ethology provide us with an ontical handle that makes such an interpretation reasonable? How might we come to interpret our fellow mammals as open to projecting their own being as being this way or that? In this segment, I interpret ethological studies in this very manner by focusing on (1) nonincidental social behaviors surrounding hunting, (2) biological death of others, (3) hierarchical relationships, and (4) toolmaking and use. I remind the reader that my aim is not to do this for each and every mammal but, more practically, to undercut Heidegger's claim of an abyss between humans and *all* of our fellow animals. Interactions among the more social animals provide particularly compelling accounts.

Orcas are large, social, apex predators who live in smallish (e.g., two to fifteen individuals) social groupings called "pods" and who hunt collectively (Grandi, de Castro, and Crespo 2012; Coscarella et al. 2015; see online videos of orca hunting). What is particularly interesting is how varied and complex their socially learned hunting techniques can be. For example, these huge marine mammals intentionally surge onto the shore to catch baby sea lions and then wiggle their way back to the open ocean with the help of wave action. They set up an ambush to capture free-swimming dolphins, whereby one orca lies in wait, while others in the pod chase the dolphins toward the waiting—and silently hiding—orca. They intentionally and collectively generate waves to knock seals off ice flows. But cleverness resides not only with the orcas. Sea lions have been observed to seek refuge under a fishing vessel and its potentially dangerous propeller in order to escape the much larger marauding orcas. If we consider their *Dasein*, we can see that orcas *are* predators/hunters and other marine mammals *are* prey/hunted—that is who they are. In other words, when thinking *Dasein*, "predator" and "prey" are existentialia, and not to be thought only as scientific descriptors that we humans assign to orcas and sea lions, respectively.

What about (existential) death, that is, death not as a biological event—which Heidegger calls "demise" for humans and "perishing" for our fellow animals—but as a way of being-toward-demise (or -perishing)? Is such thinking credible for our fellow mammals? Consider Tahlequah (also called "J35"), an orca who carried her dead calf for seventeen days and who was surrounded in a tight circle for two hours by other females in her pod as an act of seeming recognition of her distress (see King 2019b). Similar carrying of dead young, as well as other thanatological responses, have been observed in Asian elephants (Pokharel, Sharma, and Sukumar 2022; see Preston 2022), collared peccaries ("javelina"), donkeys, ferrets, common marmosets, various monkeys, and chimps (see de Waal 1996; King 2019b; Monsó 2024). As a relevant aside, one question sometimes posed is whether we should interpret these behavioral responses as examples of "long-term social distress" (i.e., to emphasize a particular scientific, reductionist accounting) or as grieving (i.e., to consider them as attunements of *Dasein*). But these two interpretations are not directly comparable, as the second interprets scientific data/observations within the frame of the animal as in *Dasein* or not, while the first interprets the animal as a biological entity with constituent building blocks and activities that characterize *all* living things as equivalent. In the case of the organism, a certain configuration of these blocks and activities simply means "distress." The shift in thinking from organism to *Dasein* requires making a new interpretation. Interpreting Tahlequah's behavior as in *Dasein* seems warranted in light of what we know about orca biology, evolutionary history, and social complexity and contingency, but still, it is our understanding of Tahlequah as Tahlequah and her particular situation that matter most. I consider this second interpretation to be warranted in light of her *Dasein*, while the first still remains a scientifically useful reductionist account.

Now let me return to projection. If Tahlequah is attuned and in *Dasein*, then being-toward-demise must also have meaning for her. Minimally, we could say that Tahlequah becomes aware that there will be no return for her calf and, therefore, Tahlequah now projects her own life to be changed. Here, we must be careful to avoid anthropocentric thinking. Even so, I consider it reasonable to think that Tahlequah realizes her future life is not to be as she had imagined before her infant died; otherwise, grieving becomes reduced to distress, that is, to a prolonged emotion which requires no awareness of past and future. Similar considerations apply to chimps (de Waal 2019b). Of course, more ethological and neuroethological research is warranted to confirm (or not) that orcas have all the biological capabilities that could enable projective behavior, but at least we should no longer deny the strong possibility of orca projection; that is, we should stop assuming Heidegger's abyss.

The third category of behaviors to be interpreted within the frame of projection is intra-group hierarchical relationships. Van Vugt and Smith (2019) (see also Smith and van Vugt 2020) distinguish two, noncoincident forms of hierarchy: (1) exerting disproportionate influence on group behavior in a decision-making hierarchy, namely, "leadership" and (2) winning access to resources and dyadic struggles over those resources, namely, "dominance status." Chimps who are α-males within their patriarchal group (de Waal 2019b), orcas who are matriarchs within their matrilines (DeWeerdt 2019), and alliances of female bonobos guided by a matriarch within their matriarchal group (Chapter 5 in de Waal 2022) are well-documented examples of leadership, although, for example, *individual* female bonobos are not necessarily dominant in both senses outlined above. (See Chapter 3 in de Waal 1996 AND Smith et al. 2023 for other examples.) In particular, leadership is not dictatorship—even α-male chimps must sometimes defer to more dominant (α-)females (de Waal 2019b). And, α-male chimps who resort to bullying others in their group rather than skillfully building coalitions—not to mention resolving others' intra-group conflicts and consoling others who are on the losing end of some happening—are less effective in maintaining their leadership (de Waal 2022: 209–10). All this points to apes and orcas having a behavioral complexity built on prior learning and sensitivity to one's situation. Interpreted within the context of *Dasein*'s projection, these various creatures come to project what leadership means to them within the context of their group; they literally project themselves as leaders. In similar ways, other less-dominant group members project other identities.

The last category of behaviors is toolmaking and use. Oftentimes discussed within the frame of cognitive abilities and presencing, toolmaking and use also have a clear projective aspect. Ontically, they enable some imagined future end. Ontologically, while they must reflect *all* of temporality's modes of being *Dasein*, their projective aspect indicates that some *Dasein*s can be tool users who not only use tools now but who also see themselves as committed to using tools in some appropriate, future situation. For example, a capuchin monkey does not automatically use a stone to crack a palm nut when in the presence of both objects; rather, even *before* the capuchin cracks a nut, she must already have projected herself as a tool user who may, or may not, choose to take advantage of the affordances. Her behavior is simply too complex to consider otherwise (see Visalberghi et al. 2015; Coghlan 2015; Fragaszy et al. 2023). For example, the capuchin may carry a stone—to be used as a hammer—to a distant location that provides a more suitable pounding surface (i.e., as an "anvil") and that provides access to cracking nuts. How else does she "see" herself while carrying the stone and before seeing the pounding surface and nut if not as a tool user? Furthermore, and consistent with *Dasein*, this "seeing" is not simply an inheritance. Rather, it emerges slowly as the capuchin observes others who are more proficient and as she strives to improve her own skill, learning for example to choose appropriately sized stones, to apply distinct amounts of pounding force when removing the outer and the inner portions of the nut, to recognize and avoid trying to crack particularly hard nuts, and many others. During this prolonged training period, she is projecting herself as becoming a better tool user. I am not saying that biology or evolution has no role here. For example, 2- to 3.5-year-old children spontaneously (that is, without prior cultural knowledge) solve certain novel tasks using tools, and "there is a close relationship between the difficulty level of these behaviors and individual discovery rates for both humans and great apes," consistent with tool use in their last common ancestor (Reindl et al. 2016).

In addition to their using stones as hammers for breaking nuts, chimps also use wooden objects which they may intentionally modify for improved functionality. We could consider chimps to be toolmakers as well. Indeed, de Waal (2016a: 84) says,

> Chimpanzees, like all apes, think before they act … [while] the capuchin monkey is a frenzied trial-and-error machine. These monkeys are hyperactive, hypermanipulative, and afraid of nothing. They try out a great variety of manipulations and possibilities, and once they discover something that works, they instantly learn from it.

While de Waal's reference to "machine" is problematic, he usefully points out important intra-primate differences that warrant thinking about (some species of?) primates as having different forms of *Dasein* (see also Visalberghi et al. 2015). Capuchins may not understand why they successfully accomplish some task, while chimps have a sense of the cause-and-effect relationship and plan accordingly (de Waal 2016a: 84–7, 257). Like capuchins, chimps must train their young to crack nuts with stones, and it takes years for them to become proficient (de Waal 2016a: 257). (See a parallel account of training young chimps in the use of sticks and twigs to "fish" for termites [de Waal 2022: 260].) Yet, the young chimps persist, even over the extended beginning when their efforts fail completely and their fingers may get smashed. Are they projecting themselves out of their temporarily deficient state and into one that matches that of their mother?

Bonobos, like chimps, readily use tools in captive and semi-captive conditions and display forward thinking in their tool use, but interestingly, bonobos do not use tools in the wild (de Waal 2022: 119–20). De Waal (2016a: 176) has proposed that a bonobo's ability to manipulate others could be considered a social tool. Like bonobos, orangutans make little or no use of (technical) tools in the wild, despite their highly dexterous tool use in captivity (de Waal 1996: 231; de Waal 2016a: 81) De Waal (2022) says, "The laboratory wins hands down on this count: observational learning covers a wide range of processes that cannot easily be disentangled in the field." If our aim is to understand the various forms of *Dasein*, then a combination of field observations and experimental interventions would seem warranted, perhaps even necessary.

Being the most intensively studied of great apes, chimps provide clear evidence of planning in their use of tools, where "planning" points to projection. For example, wild chimps carry grass stems—to be used as tools for extracting termites from their mounds—for miles until they arrive at the mounds (de Waal 2013: 205). And, in a Swedish zoo the chimp Santino threw stones, carefully cached the day before, at human onlookers. Furthermore, chimps can combine multiple tools in order to access a food reward, for example, by stacking multiple boxes to get closer to the food (de Waal 2016a: 63–4). Given the readiness with which the chimp named Sultan accomplished this task, we can say that he imagined—or projected—a better tool and then built it in order to get the reward. Finally, I note that the tools of chimps can be more complex than grass stems and stones. For example, chimps in Gabon have five-piece tool kits which enable them to hunt for honey (de Waal 2016a: 79–80). Individual tools include a pounder, a perforator, an enlarger, a collector, and swabs.

Tool use by mammals other than primates may be less common, but it is not unknown. Some bottlenose dolphins carry sponges on their snout and jaws as protection from injury while digging in the sand for prey (Mann et al. 2008). Elephants scoot boxes along a floor to a preferred location so that they can then use them to reach food otherwise out of reach (de Waal 2013: 115). And, sea otters place stones on their chest and, using their front paws, smash a shellfish against the stone in order to access the edible insides (de Waal 2016a: 88).

Now, let us switch from thinking about temporality to spatiality. For our fellow mammals, spatiality points to particular sites wherein their *Dasein* might dwell—in nearness or in farness, as Heidegger says for human *Dasein*. As *Dasein* is an ontical being who dwells, or not, there must be ontical correlates that we humans can interpret as dwelling and non-dwelling. For ontical correlates of temporality, we looked to a creature's behavior, but for spatiality, we become aware that there is also a background that enables or distorts that behavior. This background cannot be described by spatial coordinates alone; rather, it encompasses the meaning of the entire situation in such a way that the mammal's behavior is not fully understandable without it. For example, imagine the distinct behaviors that could be evoked by a mammal being (1) in an industrial slaughterhouse or a laboratory for experimentation, (2) in a displaced wild habitat (e.g., for a non-native species introduced for biocontrol purposes OR a species forced into adjoining areas due to habitat loss), (3) in one's otherwise usual habitat that has become devoid of one's own kind (e.g., through overhunting), and (4) in one's usual habitat surrounded by kin. Like *Dasein*, backgrounds too are dynamic. For example, coyotes in North America used to reside in the open spaces of the West, but now—in response to habitat alteration (e.g., transformation of prairie lands into farms) and government efforts at coyote eradication—coyotes have increased their rate of reproduction and expanded their habitat range to the entire eastern temperate and subarctic regions, including major urban areas (Flores 2016; Kays and Michelson 2018). Imagine how nearness and farness must be altered for a coyote to live in an urban space, and how much cleverness the coyote must have to accomplish this. (The same might be said for wild turkeys in the Northeastern United States.)

As another example, many mammals, including most humans, have a home space that they inhabit, and trips outside that space are typically brief and could be referred to as "visits." Even human wanderers are considered unusual *because* they stay away from their home space. But some of our fellow mammals routinely inhabit discontiguous spaces at different seasons. We call these species "migratory," and they include numerous marine mammals (e.g., orca and whales) and some terrestrial mammals (e.g., wildebeest, elephants, bighorn sheep, moose). Studies in bighorn sheep and moose demonstrate that their migratory routes are learned through experience and are culturally transmitted (Jesmer et al. 2018; Festa-Bianchet 2018). These routes and different seasonal spaces relate to successful predator avoidance, lambing, and foraging—and all are complex, meaningful aspects of their way of being. We could say that migratory animals have an awareness of space that differs from non-migratory animals in meaningful ways.

But space, even when measured scientifically at a smaller scale, can also be meaningful for the individual mammal. Consider bats, who can fly and skillfully navigate in darkness (e.g., O'Mara and Ziegler 2024). Neurological studies (Dotson and Yartsev 2021) demonstrate that

> neural activity in area CA1 [of the brain] predominantly encodes nonlocal spatial information up to meters away from the bat's present position. This spatiotemporal representation extends both forward and backward in time … and is found during both random exploration and goal-directed navigation.

Bats are aware of where they have been and where they are going. In psychological terms, they do this by constructing cognitive maps (see Toledo et al. 2020; Harten et al. 2020). In natural situations, the particular encoded spatial locations relate to what a bat already has or will have: capture and eat, mate with, avoid, explore, etc.; that is to say, they are meaningful.

As a final example, I turn to rats who were trained to play the "hide-and-seek" game with their human experimenters (Reinhold et al. 2019). The rats were fast learners for both roles—hiding and seeking. No food rewards were used. In fact, the rats displayed multiple movements and vocalizations that suggested their eagerness to play the game for itself. Importantly, these differed from those expressed when rats participate in a more coercive, operant-conditioned activity linked to a food reward. When playing hide-and-seek, the rats remembered past hiding places and were particularly quiet when hiding. The topological complexity that these experiments present to the rats is considerable, but here I emphasize that, when hiding, a rat designates one place as "for hiding" and not the other places, and that chosen hiding places vary from game to game. When seeking, the rat remembers the experimenter's past hiding locations as "previously used for hiding." Overall, I interpret this as indicating that the rat is producing his (male rats were tested) own meaning based on his understanding of existential space.

Section F. The Observer and the Observed

In making interpretations of *Dasein*, I have relied on observations from ethologists and related biologists who ply their trade using the formal scientific methodologies of their training. But at times, a few biologists have also chosen to write popular accounts of a relationship they had with a particular nonhuman animal (human/parrot: Irene Pepperberg 2009—*Alex & Me*; human/various wild birds: Bernd Heinrich 2016—*One Wild Bird at a Time*; human/fox: Catherine Raven 2021—*Fox and I*; human/owl: Carl Safina 2023—*Alfie & Me*; human/octopus: Craig Foster 2020; see also Foster 2024). Unlike standard ethological accounts where the emphasis is on "objectively" describing this or that observed animal in remove from the observer, these popular accounts explicitly include the observer, the observed, and their relationship. But a question arises: *If* the goal of the human observer is to interpret an animal's behavior in terms of *Dasein* (or some other whole-organism understanding such as consciousness, sentience, sapience), or not, how do we separate the observer and the relationship from the observed? As I see it, there is a problem here with the question itself. Except under the most sterile of laboratory conditions, observed animals always have an environment that includes other beings—living and/or inanimate—that affect the observed animal. The goal of the observer is not necessarily to eliminate these but to hold them constant enough and to acknowledge their presence in the scientific report when deemed relevant. The presence of a human observer who interacts with the observed is not so different. The observer can still explicitly interpret the observed's behavior in terms of *Dasein* (or consciousness, etc.) while acknowledging their own presence along with the particularities of their interactions. The authors already cited in this paragraph do this in one way or another, although not within the framework of *Dasein*. So, I conclude with a highly abbreviated personal account of my own friendship with Mineral, a cat who shared my residence for almost seventeen years (see List 5.1).

List 5.1. *Evidence that Mineral [a particular cat] is in* Dasein *based on the author's interpretation of their interactions*

Mutual affection:

- Mineral repetitively paws (kneads) me.
- Mineral eagerly receives my rubbings, scratchings, and pettings.

Mutual game-playing within context of Mineral as predator:

- Mineral plays "catch-and-bite" (his version) or "bait-and-switch" (my version) when I wave my index finger near his face for him to "catch" the finger (always done relatively gently by Mineral and cautiously by me!).

Mutual friendship:

- Mineral lies on or beside me while I'm napping or sleeping.
- Mineral and I walk around the apartment together.
- Mineral sits on my lap during breakfast and while I'm reading.
- Mineral and I frequently talk to each other.

Mineral's expression of care:

- Mineral licks my hair when I'm lying down.
- Mineral paws my face when he wants an activity (e.g., petting) to continue.
- Mineral waits by the front door for me when I am away during the day.
- Mineral is visibly affected when I have been away for longer periods.

My acknowledgment of our reciprocal relationship:

- I carry Mineral around our apartment in a box, showing him "interesting" sites.
- I brush Mineral's teeth, feed him, and keep his litter clean.

Affection, game-playing, predation, friendship, care. In this book, I have tried to show that these are expressions of *Dasein*'s way of being through existence with its three aspects of temporality and spatiality. Within the interpretative frame of making sense of our world together, both Mineral and I are—in our different ways—there. Of course, scientific understandings of cat behavior could also be used to bolster my claim that cats are in *Dasein* (see Bradshaw 2014; Montague et al. 2014; Bradshaw 2017; Schötz 2018; Grimm 2019; Losos 2023).

AFTERWORD

In the Preface, I said that thinking through my own confusion over human tolerance of the continuing and now massive human-caused devastation of our fellow animals was my principal motivation in undertaking this writing project. Now, having developed my claim of animal *Dasein*, we might ask whether this can unsettle any understanding of the devastation. This is a large topic, worthy of its own book, but I would feel remiss if I failed to make a further connection (i.e., beyond what I said in "*Dasein*'s Forms and Biological Research" [Chapter 5, Section B]) to one way of human thinking and acting to which I have been deeply committed for many decades, namely, biological research. In his late-period book entitled *What Is Called Thinking?*, Heidegger (1976: 29) writes about devastation as something that "blocks all future growth." In my view, the increasing prominence of reductive biological research over the past 150 years has contributed and continues to contribute to the ontological and ontical devastation in which we now live. Of course, scientific research itself occurs within a broader, enabling social context. Nevertheless, reductive biological research has directly fostered the public's and the researchers' view of animals, including *individual* animals, as collections of mechanisms and as model systems rather than as the animals they are—that is, as in *Dasein*. I am not suggesting that reductive biological research can have no role in fostering animal-*Dasein* thinking. My interpretative focus on Panksepp's neuroethological research on the rat in Chapter 5 illustrates a largely compatible approach. However, many reductive biological studies—particularly in biotechnology—cannot illuminate an understanding of animal *Dasein* and can lead to great devastation. Likewise, talk about transforming biology into chemistry pushes away from any interest in animal *Dasein*.

Animal *Dasein* with its related forms and individual selves who meaningfully exist in their world is, in my view, one appropriate way to think being's gifting of a world to us and to others. Ethologists and evolutionary biologists are particularly well situated to express this gifting through their science. Why we would put in peril animal-*Dasein*'s world(s) remains unfathomable to me, but I now believe that the ontical challenges presented by the devastation of animal *Dasein* will remain unless we rethink Heidegger's question of the meaning of being.

NOTES

Preface

1 "Our fellow animals" will be my term of preference when referring to "non-human animals," "the other animals," or simply "animals," although sometimes I do use these latter terms when doing otherwise would make the sentence sound awkward.

2 I am not aware that Heidegger makes a decisive distinction between "entities" and "beings." Humans and our fellow animals are generally referred to as beings. An "entity" is anything that is—beings, natural objects and processes, imaginary and real thoughts, thoughts about thoughts, etc.

3 "Ontical" pertains to entities as entities, including their specific properties. The scientific investigations of humans and our fellow animals are ontical. "Ontological" (as Heidegger understands this word) pertains to being, including the being of entities and the structure of that being. Heidegger thematized human being in *Being and Time* and animal being in *Fundamental Concepts of Metaphysics*.

4 In fairness to Heidegger, I note that biologists' understanding of the term "instinct" has changed since *Fundamental Concepts of Metaphysics* was written, and indeed the term itself is now little used by biologists. This is because many so-called instinctual responses in animals turn out to be invariant only under some background conditions of scientific observation. Thus, formerly instinctual responses have been shown to express degrees of flexibility that appear continuous with more complex behaviors, including those of humans. Heidegger was aware of some of this complexity, but it did not alter his ontical claim of pervasive animal instinct.A word closely related in meaning to "instinct" is "innate." I avoid discussion of this word as it is not used in my translation of *FCM*, while "instinct" is frequently used.

Introduction

1 Shared derived characters—also called "synapomorphies"—are those characters that delimit an entire taxonomic group. They arise at the time of the group's origin and are found in all of the group's (later) representatives, although sometimes in modified forms (Hennig 1979). Examples of shared derived characters for the entire group of animals ("Animalia") are the development from a blastula, the presence of HOX (homeotic) genes, and multicellularity. Although multicellularity is found in many species outside animals, it is still uniquely derived for animals because animals evolved from a single-cell ancestor. In other words, multicellularity has evolved independently in multiple lineages, and it is therefore a separate shared derived character for other lineages. Shared derived characters can be secondarily lost within a lineage, and that loss itself thereby becomes a shared derived character for the sublineage, for example, the loss of flight in certain birds.

You might wonder whether and, if so, how evolutionary groupings—and evolutionary studies generally—relate to Heidegger's and to my understandings of animal being, but for that you must be patient and wait until Chapter 3!

Chapter 1

1 For Heidegger, the word "fundamental" is closely related to "ontological," a central concept in his writing that refers to an understanding of being rather than beings.

2 Heidegger states that inauthenticity should not be viewed negatively or as a value judgment but as the possibility of our everyday way of being. Ontically, this is expressed in social cohesiveness and politeness but also in "not challenging the status quo" and in "being unwilling to confront our own failures." A broad reading of Heidegger makes it difficult not to assign negative valence to inauthenticity.

3 Briefly and as described in the second division of *Being and Time*, authenticity in human *Dasein* emerges from inauthenticity when *Dasein* "accepts" her local situation as her very own, and couples this with the projecting of a futurally oriented for-the-sake-of-which. This projection can be either novel or traditional, but it must be "accepted" by an individual *Dasein* as her own and not, for example, merely "absorbed" from the surrounding culture. Heidegger is clear that authenticity can never be a fully willful action by someone.

4 Heidegger asks this question (*FCM:* 6) to remind us that what should be of concern is our own *Dasein*, and not the acquisition of a third-person explication of some matter.

5 Related to this point, Heidegger says, "Philosophy is the opposite of all comfort and assurance" (*FCM:* 19).

6 I have chosen not to translate *Dasein* from the German.

7 As examples of prevailing, which are not to be thought in biological terms, Heidegger mentions "procreation, birth, childhood, maturing, aging, [and] death" (*FCM:* 26). John Haugeland (2013) helpfully interprets: procreation or trust in possibilities; birth or possibilities inherited; childhood or discovering *Dasein*'s own possibilities; maturing or searching for new possibilities; aging or accepting the diminution of possibilities; death or the ending of all possibilities.

8 In brief, the distinction between "attunement" and "profound attunement" is that while the former identifies *Dasein*'s "mood"—as described in *Being and Time*—the latter grabs the whole of *Dasein*'s existence and provides her basic bearings.

9 The creatures mentioned by Heidegger in *FCM* and his other writings are now classified in the following manner:

1. LIVING BUT NOT ANIMALS: amoeba, infusoria
2. ANIMALS BUT NOT VERTEBRATES: bee, glow worm, worm, fly, moth, sea urchin, wood worm
3. VERTEBRATES BUT NOT MAMMALS: lizard, chaffinch, "bird of prey," fish, great tit, woodpecker
4. MAMMALS BUT NOT PRIMATES: dog, domestic pets, mole, squirrel
5. PRIMATES BUT NOT HUMANS: (none)
6. HUMANS: *Homo sapiens.*

10 Heidegger does say that humans can "animate" the stone through myth and art, giving rise to a quite different kind of accessibility ("truth"). But, for the purposes of *FCM*, Heidegger wants to stay within the scientific and metaphysical traditions, and so does not further elaborate.

11 In terms of ontogeny, many animals go through a single-cell stage, in particular, when the sperm and egg fuse. So, any claim that animals are intrinsically multicellular would be to display an adult bias. My point is that a commitment to organismic holism is separate from the issue of multicellularity. Heidegger's claim for animals that "The unit of life is not the cell" borders on conflating holism and multicellularity.

12 Heidegger's holistic thinking about the subservience of organs to the organism is particularly well suited to eyes, where the optic nerve directly connects the external eye to the brain. As a more complexifying example, the pancreas can produce its digestive enzymes even when this organ is cultured outside the body. Such enzyme production is—according to biologists—the pancreas's function, like seeing is for eyes. But for Heidegger, enzyme production cannot meaningfully be described as a function. Rather, the pancreas is subservient to the capability for digestion, which is characteristic of the organism. The pancreas results from the digestive powers of the organism, and not the reverse, despite its (partial) autonomy.

13 Darwin remained receptive to the inheritance of acquired characteristics, and, even today, the evolutionary significance of so-called Lamarckian evolution remains uncertain. Well-established pathways ("mechanisms") of heritable, population-level genetic variation—besides Darwinian natural selection—include genetic drift and gene drive.

Chapter 2

1 *Seyn* (be-ing) has also been translated as "beyng" and "historical being."

2 The plants and animals mentioned in this essay are: thrush, forest, blue wild game, corn flower sheaf, flowers, animal, blackbird, rushes, poppy, fish and game, snakes, hyacinth, birds, myrtle, oaks, elms, reeds, blossoms, fir trees, fishes, and myrtle. References to these plants and animals are directly taken from Heidegger's quotations of Trakl's poems.

Chapter 3

1 Here, I offer two comments on Heidegger's cultural background, despite also knowing that linking such information to his philosophical thought would be, according to Heidegger, to turn away from philosophy and toward naturalism, e.g., psychology and sociology. First, Heidegger's early cultural and religious upbringing, including briefly becoming a Catholic priest-in-training, likely played a formative role in his later thinking (Heidegger 2010). Second, according to noted evolutionist and historian Ernst Mayr (1982), Germany (probably meaning its scientists and other intellectuals) in Heidegger's time (and earlier) largely accepted evolutionary thinking but largely rejected Darwin's 1859 theory of natural selection (see Darwin 2006). However, absent a "mechanism" such as natural selection, evolution as a science loses its explanatory power and even becomes open to a fundamentalist religious interpretation. Mayr suggests that the popularization of Darwinism by the famous German biologist Ernst Haeckel, particularly in the context of Haeckel's avowed atheism (Haeckel 1905), led to non-science-based counterattacks on Darwin's theory of natural selection as fostering a virulent form of materialism, and immorality. It is clear in *FCM* that natural selection, not evolution per se, is Heidegger's main target as concerns Darwin (e.g., *FCM*: 277–8), so Heidegger could be considered to be in the mainstream of German intellectuals of his time, although not those in other countries where the major scientific advances in the new evolutionary biology were ongoing, notably in England and the United States. I wonder how Heidegger would

have responded to the now-large literature on altruism in evolution (e.g., Nowak and Highfield 2011; E.O. Wilson 2012; D.S. Wilson 2015).

2 Here is how I would summarize a current understanding of Darwinian evolution via natural selection:

- Precondition #1: There is a target population of self-maintaining and self-reproducing organisms.
- Precondition #2: Across this population, the individual organisms display phenotypic variation.
- Precondition #3: The phenotypic variation arises from genetic mutation and is heritable in those variants that are able to leave offspring.
- Precondition #4: Mutations are non-directed (i.e., stochastic) with respect to the organisms' resulting phenotypes.
- Precondition #5: Organisms within the population compete for resources in the external environment.
- Precondition #6: Organisms with different phenotypes may have different degrees of success in competing for resources.
- Precondition #7: More successful phenotypes will, on average and all other things equal, have relatively more offspring, and, correspondingly, less successful phenotypes will have relatively fewer offspring.
- Outcome: More successful phenotypes will—over the ensuing generations—tend to increase in abundance relative to less successful phenotypes.

The above statements taken together constitute what I see as the current understanding of Darwinian evolution via natural selection. In such Darwinian evolution, phenotypic changes at a multi-generational timescale follow from all of the preconditions—or, taken together, "the state"—of the entire system (i.e., population + its environment = "the natural world") rather than from a direct selection process, e.g., sometimes confusingly articulated as the environment "selecting" or "filtering" the population's phenotypes (see also Thompson 2007). Relatedly, natural selection is oftentimes referred to as a "mechanism," but if so, it is a higher-level sort of mechanism than that discussed by functional, reductive biologists.

3 A complexification in the standard understanding of Darwinian natural selection: the external resources of an evolving population (see previous footnote) can also vary. When these external resources are non-target organisms (e.g., consider predator and prey or herbivore and plant as target and non-target), they too can undergo Darwinian evolution, leading to further natural selection among target organisms. Nonliving resources may also undergo changes (e.g., consider changes in the climate), and while these changes do not directly follow the path of Darwinian evolution, they may also lead to further natural selection of target organisms.The basic problem for evolutionary biologists in light of this complexification is that the target population and the environment are never completely separable. Even so, there are circumstances when changes in the target population and its environment are sufficiently disconnected to make accurate approximations (see Lewontin 2000). Knowing when these approximations are more or less accurate is part of the reason why evolutionary biology is a probabilistic science, and not a deterministic one like classical physics.Heidegger's understanding, like that of many scientists even today, of evolution and natural selection was not as holistic-thinking-friendly as I think it should have been. I believe Heidegger's negative critique of evolution and natural selection no longer has warrant, which is why I have explained all this in some detail.

4 Heidegger's understanding of "interpretation" is distinct from "inference." The latter is based on formal logical reasoning, which Heidegger says cannot directly apply to an understanding of being, in part, because logical inference gives priority to entities. Even so, inferences—along with "observations"—can provide a useful background when making reasoned interpretations.

5 In translating *Historie* as *"history,"* and *Geschichte* as *history*, I am adopting the convention proposed by Parvis Emad and Thomas Kalary in the Translators' Foreword (p. xxxii) of Heidegger (2016c)'s *Mindfulness.*

6 Here, it is important to note that the ways that bacteria (and archaea) live in their natural environment can be much more diverse and complex than typically described by scientists in their highly controlled laboratory experiments. For example, bacteria can have multicellular forms with cell-specific functions, as well as systems of sensing and communication. For a sensitive and accessible view of natural bacterial life, see Chimileski and Kolter (2017).

Chapter 4

1 Bees in the studies cited are either the domestic honey bee or one of a few species of bumble bees. There are approximately 20,000 other species of bees with widely divergent life histories.

2 One could argue that any intervention—even in the affordances of an animal—contravenes holism. But taken to this extreme, all experimentation becomes impossible. I agree with Lewontin (2000), who says that the aim of science is to identify those articulations of the world that enlighten our understanding without fundamentally denying the world's holism. It is by varying the affordances where we have the most to learn about an animal's way of being. Articulating within the body of the animal can be ontologically enlightening only when the observations are interpreted in the context of the whole organism. Making a fuller argument for this claim is beyond the scope of this book.

3 After I completed a draft of this manuscript, the biologist Robert D. Seeley published (in April 2024) a wonderful book (*Piping Hot Bees and Boisterous Buzz-Runners*) chronicling his luminous and long career as an investigator of bee behavior within the broader community of bee behavioralists. The book is rich in documenting numerous fascinating behaviors beyond the foraging behavior of the female worker bees that I have emphasized. He also documents information flow within the colony via various mechanical signals (e.g., tremble dance, worker piping) and chemical signals. Although his focus is on behavior of groups of bees rather than of individuals, there is much that could be said about individual bees as well and that could be extended to an interpretation of bee *Dasein*. Seeley says, "But in the early 1970s … I began to appreciate the astonishing behavioral complexity of worker honey bees, especially when they function as nest-site scouts. Now, having spent five decades studying how these bees function in various roles—nectar forager, water collector, resin collector, and nest-site scout—I see clearly, and I respect deeply, the behavioral sophistication and cognitive abilities of these little beings" (p. 265). In light of Seeley's book, I humbly submit that what I have chosen to write about in this chapter documents but a small fraction of the wondrous life of a bee. Seeley's book (p. 202) also made me aware of "Tinbergen's four questions" as a categorization of different ways of studying

animal behavior. They are: (1) "How does it work?" (i.e., a question of function), (2) "How does it develop in the individual's lifetime?" (i.e., a question of ontogeny or development), (3) "Why is it beneficial?" (i.e., a question of adaptation), and (4) "What is it's evolutionary history?" (i.e., a question of phylogeny).

4 In *Being and Time* (pp. 456–7) Heidegger says, "All Dasein's behaviour is to be Interpreted in terms of its Being—that is, in terms of temporality."

5 Donald Griffin (1981) makes the point that the waggle dance directs recruits only to a general area, and that the exact location must be identified by the recruits' own sensory perception, e.g., by following the scent that the dancing bee picked up from the site of the flowers.

6 The mirror (or mark) test is the classic test of self-awareness in humans, and it is based on mirror self-recognition. For example, a human child who is familiar with mirrors has a visual mark applied to her forehead in a position where she is unable to see that mark without a mirror. Then, she is given the opportunity to look at herself in a mirror. If she spends an unusual amount of time examining the mark in the mirror, and if she touches the mark on herself (not the mirror) or tries to rub it off, then she is judged to be aware that the mark is both foreign and on her, i.e., she is self-aware of the mark. A key assumption of the test is that—with the aid of a mirror—a self-aware child recognizes the mark on her body as foreign and, therefore, as warranting her increased attention.

7 Chittka uses the term "anxiety" rather than "fear," but I follow Heidegger when I say that fear has a definite object, while anxiety is not so clearly connected to an object.

8 Learning by imitating the behavior of others is a trait that humans frequently exploit as well.

9 Bodily self-awareness has also been experimentally documented in the non-bilaterian Caribbean box jellyfish (phylum Cnidaria) (Bielecki et al. 2024).

10 "Clearing" and "illuminating" illustrate the centrality of metaphorical language in Heidegger's attempt to speak the unspeakable, namely, being.

Chapter 5

1 One recent estimate of terrestrial arthropod-species number is 7 million, including 5.5 million insects and 1.5 million beetles (Stork 2018). Currently, only about 1 million insect species have been named.

2 I say "relatively" because moths do display group aspects in their mating and migratory behaviors.

3 Strictly speaking, agriculture is not limited to immediate food production. Consider that humans raise farm animals for work, recreation, medicinal production (e.g., vaccines and antibodies), biomedical research, and consumer products (e.g., sheep's wool). Consider the growing of trees for paper products, building materials, and conservation (e.g., as windbreaks). Consider the growing of fungi for pest control and medicinal production. Even so, the current discussion is limited to food production.

4 My biological discussion of agriculture is largely drawn from the excellent multi-authored book entitled *The Convergent Evolution of Agriculture in Humans and Insects* (Schultz, Gawne, and Peregrine 2021). In their introduction, the editors define agriculture as "Cultivation on a large scale in which the farmer has become obligately dependent on the cultivated species for nutrition."

5 Embryonic bilateral symmetry evolved early in evolution of animals, but still after their origin. Bilaterian animals exclude the deep lineages of sponges and cnidarians, for example. The deepest temporal divergence within bilaterian animals is between protostomes (e.g., platyhelminths, nematodes, mollusks, arthropods, + other less-speciose groups) and deuterostomes (e.g., echinoderms, hemichordates, chordates such as vertebrates, + other groups).

6 For example, Jacobus Boomsma (2021: 77) reductively says with reference to fungus-farming ants, "… all individual agency in the sense of independent decision-making has disappeared out of their family lives no matter if one is a queen, a worker, or a soldier. Lifetime unmated workers are as committed to their family as somatic cells are to a metazoan body."

7 To avoid verbal confusion, I distinguish between sensory injury—where injury is to be broadly understood—and pain, which is the awareness of that injury. Doing this separates the scientific observation from its interpretation.

8 Consider the non-bilaterian box jellyfish (phylum Cnidaria) who can learn to avoid obstacles, even though its simple nervous system, connected as it is to image-forming eyes, is not organized into a centralized brain (Bielecki et al. 2023, 2024). This should caution us against being overly "brain centric," just like I previously cautioned against "brain-size centrism" in my discussion of social bees (see Chapter 4). Regardless, the jellyfish avoidance behavior is likely based on associative—not cognitive—learning, and while this is not inconsistent with awareness, it does not establish it.

9 For simplicity's sake, I leave out many important details. For example, the crabs display a bias for leftward movement independent of any shocking, so when I say that the crabs show "no preference," this already takes into account—or subtracts out from the determination—their own biased directional movement. Another important point is that the described changes in behavior are probabilistic; that is, not all crabs respond, or respond in the same manner.

10 A later study by the same coauthors (Magee and Elwood 2016) used a modified experimental design and arrived at partially distinct conclusions, emphasizing the challenge of generalizing from specific ethological experiments.

11 As a placeholder term, "cephalopod *Dasein*" is not meant to imply that all individuals of cephalopod species have the same form of *Dasein* or even that they are in *Dasein*. I note that there are 650–700 extant species of cephalopods (octopus, squid, cuttlefish), of which perhaps 300 are octopuses.

12 Scheel (2023) poignantly analogizes this octopus holding its injured arm to a human sucking its thumb after being struck with a hammer.

13 Dwelling in hermit crabs can also be affected by human trash (see Jagiello, Dylewski, and Szulkin 2024).

14 One possibility in the near future is that artificial intelligence will enable cyborg *Dasein*. Obviously, this important topic is beyond the aims of this book.

15 I say "irritants" because that is how they would be interpreted *if* they had been injected into the lips of humans and other terrestrial vertebrates. This interpretation also draws strength from how biologists interpret the irritants' chemical reactivities when injected into living cells.

16 Frans de Waal (2016b, 2019a) also makes the reasonable point that interpreting the mirror self-recognition test as a simple binary (i.e., pass/fail) is likely to be an oversimplification; that is, it fails to do justice to the various forms of self-awareness.

17 Here, "region" can be interpreted as the where-to and what-for of a world's meaningfulness (see "Spatiality [*Räumlichkeit*]" entry on pp. 199–202 in Dahlstrom 2013).

18 "Deseverant" is one translation of *Dasein*'s ability to bring into nearness what is remote.

19 See also Ben-Ami Bartal (2024) for a more recent broad assessment of affective and cognitive capacities in rats. Interpreting the import of this article extends beyond beenness to the other aspects of temporality, but that task I leave to the reader.

20 "Brain/Mind" and "Mind/Brain" are meant to indicate a unity between the old terms "brain" and "mind." "Brain/Mind" emphasizes structure, while "Mind/Brain" emphasizes mental activity.

REFERENCES

Ackerman, Jennifer (2016). *The Genius of Birds* (Kindle ebook [www.amazon.com; downloaded October 20, 2016]). Penguin Press. New York.

Ackerman, Jennifer (2020). *The Bird Way. A New Look at How Birds Talk, Work, Play, Parent, and Think*. Penguin Press. New York.

Agamben, Giorgio (2004). *The Open. Man and Animal* (Kevin Attell, Trans.). Stanford University Press. Stanford, CA.

Akinrinade, Ibukun, Kyriacos Kareklas, Magda C. Teles, Thais K. Reis, Michael Gliksberg, Giovanni Petri, Gil Levkowitz, and Rui F. Oliveira (2023). "Evolutionarily conserved role of oxytocin in social fear contagion in zebrafish." *Science, 379*(6638), 1232–7.

Allen, Timothy A., and Norbert J. Fortin (2013). "The evolution of episodic memory." *Proc. Natl. Acad. Sci. U.S.A., 110*(suppl. 2), 10379–86.

Altschul, Drew M., William D. Hopkins, Elizabeth S. Herrelko, Miho Inoue-Murayama, Tetsuro Matsuzawa, James E. King, Stephen R. Ross, and Alexander Weiss (2018). "Personality links with lifespan in chimpanzees." *eLife, 7*, e33781.

Amodio, Piero, Markus Boeckle, Alexandra K. Schnell, Ljerka Ostojić, Graziano Fiorito, and Nicola S. Clayton (2019). "Grow smart and die young: Why did cephalopods evolve intelligence?" *Trends in Ecology and Evolution, 34*(1), 45–56.

Andrews, Kristin, and Jacob Beck (Eds.). (2018). *The Routledge Handbook of Philosophy of Animal Minds*. Routledge. Abingdon & New York.

Arnon, Inbal, Simon Kirby, Jenny A. Allen, Claire Garrigue, Emma L. Carroll, and Ellen C. Garland (2025). "Whale song shows language-like statistical structure." *Science, 387*(6734), 649–53.

Audet, Jean-Nicolas, Mélanie Couture, and Erich D. Jarvis (2023). "Songbird species that display more-complex vocal learning are better problem-solvers and have larger brains." *Science, 381*(6663), 1170–5.

Bakker, Karen (2022). "Phytoacoustics." *Natural History Magazine, 130*(10), 30–5.

Balcombe, Jonathan (2016a). "Einstein of the sea." *Scientific American, 314*(6), 66–9.

Balcombe, Jonathan (2016b). "Sex lives of fish." *Natural History Magazine, 124*(8), 22–7.

Balcombe, Jonathan (2016c). *What a Fish Knows. The Inner Lives of Our Underwater Cousins* (Kindle ebook [www.amazon.com; downloaded October 20, 2016]). Scientific American / Farrar, Straus and Giroux. New York.

Balter, Michael (2013). "Can animals envision the future? Scientists spar over new data." *Science, 340*(6135), 909.

Barbaras, Renaud (2021). *Introduction to a Phenomenology of Life* (Leonard Lawlor, Trans.). Indiana University Press. Bloomington, IN.

Barry, James P., Steven Y. Litvin, Andrew DeVogelaere, David W. Caress, Chris F. Lovera, Amanda S. Kahn, Erica J. Burton, Chad King, Jennifer B. Paduan, C. Geoffrey Wheat, Fanny Girard, Sebastian Sudek, Anne M. Hartwell, Alana D. Sherman, Paul R. McGill, Aaron Schnittger, Janet R. Voight, and Eric J. Martin (2023). "Abyssal hydrothermal springs—Cryptic incubators for brooding octopus." *Science Advances, 9, eadg3247*, 1–13.

Bastos, Amalia P.M., and Alex H. Taylor (2020a). "Addendum: Kea show three signatures of domain-general statistical inference." *Nature Communications, 11*(1), 2906.

Bastos, A.P.M., and A.H. Taylor (2020b). "Kea show three signatures of domain-general statistical inference." *Nature Communications, 11*(1), 828.

Bastos, Amalia P.M., Zachary N. Houghton, Lucas Naranjo, and Federico Rossano (2024). "Soundboard-trained dogs produce non-accidental, non-random and non-imitative two button combinations." *Scientific Reports, 14*(1), 28771.

Beauchamp, Tom L., and R.G. Frey (Eds.). (2011). *The Oxford Handbook of Animal Ethics*. Oxford University Press. Oxford.

Belgrano, Mateo (2021). "Access to being: Elitism in M. Heidegger's philosophy." *Revista Portuguesa de Filosophia, 77*(4), 1481–98.

Ben-Ami Bartal, Inbal (2024). "The complex affective and cognitive capacities of rats." *Science, 385*(6715), 1298–305.

Ben-Haim, Moshe Shay, Olga Dal Monte, Nicholas A. Fagan, Yarrow Dunham, Ran R. Rassin, Steve W.C. Chang, and Laurie R. Santos (2021). "Disentangling perceptual awareness from nonconscious processing in rhesus monkeys (Macaca mulatta)." *Proc. Natl. Acad. Sci. U.S.A., 118*(15), e2017543118.

Bidel, Flavie, Natalie C. Bennett, and Trevor J. Wardill (2022). "*Octopus bimaculoides*' arm recruitment and use during visually evoked prey capture." *Current Biology, 32*(21), 4727–33.e3.

Bielecki, Jan, Sophie Katrine Dam Nielsen, Gösta Nachman, and Anders Garm (2023). "Associative learning in the box jellyfish *Tripedalia cystophora*." *Current Biology, 33*(19), P4150–9.E5.

Bielecki, Jan, Tom Birkoben, Maximiliane Noll, Jan-Frederik Freiberg, Peer Wulff, Heinrich Terlau, and Hermann Kohlstedt (2024). "Pattern recognition in the box jellyfish rhopalial nervous system mimicked by an ensemble of pulsed coupled oscillators." *In*: Martin Ziegler, Thomas Mussenbrock, and Hermann Kohlstedt (Eds.), *Bio-Inspired Information Pathways. From Neuroscience to Neurotronics (Volume 16)* (pp. 335–55). Springer Nature Switzerland AG. Cham, Switzerland.

Blattner, William (2005). "Temporality." *In*: Hubert L. Dreyfus, and Mark A. Wrathall (Eds.), *A Companion to Heidegger* (pp. 311–24). Blackwell. Malden, MA.

Blattner, William (2006). *Heidegger's Being and Time*. Bloomsbury. London & New York.

Boeckle, M., M. Schiestl, A. Frohnwieser, R. Gruber, R. Miller, T. Suddendorf, R.D. Gray, and N.S. Clayton (2020). "New Caledonian crows plan for specific future tool use." *Proc. Royal Soc. B, 287*(1938), 20201490.

Bonta, Mark, Robert Gosford, Dick Eussen, Nathan Ferguson, Erana Loveless, and Maxwell Witwe (2017). "Intentional fire- spreading by 'Firehawk' raptors in Northern Australia." *Journal of Ethnobiology, 37*(4), 700–18.

Boomsma, Jacobus J. (2021). "Lifetime commitment between social insect families and their fungal cultivars complicates comparisons with human farming." *In*: Ted R. Schultz, Richard Gawne, and Peter N. Peregrine (Eds.), *The Convergent Evolution of Agriculture in Humans and Insects* (pp. 73–85). The MIT Press. Cambridge, MA, & London.

Bowers, Jessica, Tahirah Nimi, Jack Wilson, Shannon Wagner, Amarie Dragoş, and Vinoth Sittaramane (2020). "Evidence of memory in the juvenile dwarf cuttlefish *Sepia bandensis*." *Learning and Behavior, 48*(4), 420–31.

Bradshaw, G.A. (2017). *Carnivore Minds. Who These Fearsome Animals Really Are*. Yale University Press. New Haven, CT, & London.

Bradshaw, John (2014). *Cat Sense. How the New Feline Science Can Make You a Better Friend to Your Pet*. Basic Books. New York.

Brandom, Robert B. (2000). *Articulating Reasons. An Introduction to Inferentialism.* Harvard University Press. Cambridge, MA, & London.

Brandom, Robert B. (2011). *Perspectives on Pragmatism. Classical, Recent, and Contemporary.* Harvard University Press. Cambridge, MA, & London.

Brandom, Robert B. (2015). *From Empiricism to Expressivism. Brandom Reads Sellars.* Harvard University Press. Cambridge, MA, & London.

Bridges, Alice, HaDi MaBouDi, Olga Procenko, Charlotte Lockwood, Yaseen Mohammed, Amelia Kowalewska, José Eric Romero González, Joseph L. Woodgate, and Lars Chittka (2023). "Bumblebees acquire alternative puzzle- box solutions via social learning." *PLoS Biology, 21*(3), e3002019.

Bridges, Alice D., and Lars Chittka (2023). "Escaping anthropocentrism in the study of non-human culture: Comment on 'Blind alleys and fruitful pathways in the comparative study of cultural cognition' by Andrew Whiten." *Physics of Life Reviews, 44*, 267–9.

Brooker, Rohan M., Jordan M. Casey, Zara-Louise Cowan, Tiffany L. Sih, Danielle L. Dixson, Andrea Manica, and William E. Feeney (2020). "Domestication via the commensal pathway in a fish–invertebrate mutualism." *Nature Communications, 11*, 6253.

Brosnan, Sarah F., and Frans B.M. de Waal (2003). "Monkeys reject unequal pay." *Nature, 425*(6955), 297–9.

Brown, Culum (2001). "Familiarity with the test environment improves escape responses in the spotted rainbowfish, *Melanotaenia duboulayi*." *Animal Cognition, 4*(2), 109–13.

Brown, Culum (2015). "Fish intelligence, sentience and ethics." *Animal Cognition, 18*(1), 1–17.

Bruck, Jason N. (2013). "Decades-long social memory in bottlenose dolphins." *Proc. Royal Soc. B, 280*(1768), 20131726.

Brucks, Désirée, and Auguste M.P. von Bayern (2020). "Parrots voluntarily help each other to obtain food rewards." *Current Biology, 30*(2), 292–7.

Brucks, Désirée, Matthew Petelle, Cecilia Baldoni, Anastasia Krasheninnikova, Eleonora Rovegno, and Auguste M.P. von Bayern (2022). "Intra- and interspecific variation in self-control capacities of parrots in a delay of gratification task." *Animal Cognition, 25*(2), 473–91.

Brusatte, Stephen (2017). "Taking Wing. A remarkable fossil record of the dinosaurs that led to birds reveals how evolution produces entirely new kinds of organisms." *Scientific American, 316*(1), 49–55.

Buchanan, Brett (2008). *Onto-Ethologies. The Animal Environments of Uexküll, Heidegger, Merleau-Ponty, and Deleuze.* SUNY Press. Albany, NY.

Bugnyar, Thomas, Stephan A. Reber, and Cameron Buckner (2016). "Ravens attribute visual access to unseen competitors." *Nature Communications, 7*, 10506.

Burghardt, Gordon M. (2005). *The Genesis of Animal Play. Testing the Limits.* The MIT Press. Cambridge, MA, & London.

Burkett, J.P., E. Andari, Z.V. Johnson, D.C. Curry, F.B.M. de Waal, and L.J. Young (2016). "Oxytocin-dependent consolation behavior in rodents." *Science, 351*(6271), 375–8.

Buttelmann, David, Frances Buttelmann, Malinda Carpenter, Josep Call, and Michael Tomasello (2017). "Great apes distinguish true from false beliefs in an interactive helping task." *PLoS ONE, 12*(4), e0173793.

Calarco, Matthew (2004). "Heidegger's Zoontology." *In*: M. Calarco, and P. Atterton (Eds.), *Animal Philosophy. Essential Readings in Continental Thought* (pp. 18–30). Continuum. London & New York.

Calarco, Matthew (2008). *Zoographies. The Question of the Animal from Heidegger to Derrida*. Columbia University Press. New York.

Calarco, Matthew (2015). *Thinking Through Animals. Identity, Difference, Indistinction*. Stanford University Press. Stanford, CA.

Calarco, Matthew (2024). *The Three Ethologies. A Positive Vision for Rebuilding Human-Animal Relationships*. University of Chicago Press. Chicago, IL, & London.

Calvo, Paco (2022). *Planta Sapiens. Unmasking Plant Intelligence*. The Bridge Street Press.

Capobianco, Richard (2021). "Presencing (*Anwesen*)." *In*: Mark A. Wrathall (Ed.), *The Cambridge Heidegger Lexicon* (pp. 603–5). Cambridge University Press. Cambridge & London.

Chamberlain, Lesley (2013). "Heidegger as a Post-Darwinian Philosopher." *Philosophy*, *88*(345), 387–410.

Chang, Weipang, and Melina E. Hale (2023). "Mechanosensitive signal transmission in the arms and the nerve ring, an interarm connective, of *Octopus bimaculoides*." *iScience*, *26*(5), 106722.

Chase, Ava R. (2001). "Music discriminations by carp (*Cyprinus carpio*)." *Animal Learning and Behavior*, *29*(4), 336–53.

Chen, Jiani, Yuqi Zou, Yue-Hua Sun, and Carel ten Cate (2019). "Problem-solving males become more attractive to female budgerigars." *Science*, *363*(6423), 166–7.

Chen, Zhanqi, Richard T. Corlett, Xiaoguo Jiao, Sheng-Jie Liu, Tristan Charles-Dominique, Shichang Zhang, Huan Li, Ren Lai, Chengbo Long, and Rui-Chang Quan (2018). "Prolonged milk provisioning in jumping spider." *Science*, *362*(6418), 1052–5.

Chimileski, Scott, and Roberto Kolter (2017). *Life at the Edge of Sight. A Photographic Exploration of the Microbial World*. Belknap Press. Cambridge & London.

Chittka, Lars (2022a, July 29). "The consciousness of bees." *Washington Post*.

Chittka, Lars (2022b). *The Mind of a Bee*. Princeton University Press. Princeton, NJ, & Oxford.

Chittka, Lars (2022c). "'Personality' differences between bees." *Natural History Magazine*, *130*(8), 16–23.

Cissewski, Julia, and Lydia V. Luncz (2021). "Symbolic use in the wild chimpanzee gestural communication?: A theoretical framework." *Frontiers in Psychology*, *12*, 718414.

Clark, Stephen R.L. (2011). "Animals in classical and late antique philosophy." *In*: Tom L. Beauchamp, and R.G. Fry (Eds.), *The Oxford Handbook of Animal Ethics* (pp. 35–60). Oxford University Press. New York.

Clayton, Nicola S., and Anthony Dickinson (1998). "Episodic-like memory during cache recovery by scrub jays." *Nature*, *395*(6699), 272–4.

Clayton, Nicola S., Kara Shirley Yu, and Anthony Dickinson (2001). "Scrub Jays (*Aphelocoma coerulescens*) form integrated memories of the multiple features of caching episodes." *Journal of Experimental Psychology: Animal Behavior Processes*, *27*(1), 17–29.

Coghlan, Andy (2015). "Smart monkeys perfect tool use to crack nuts." *New Scientist*, *226*(3020), 11.

Collet, Julien, Takao Sasaki, and Dora Biro (2021). "Pigeons retain partial memories of homing paths years after learning them individually, collectively or culturally." *Proc. Royal Soc. B*, *288*(1963), 20212110.

Colony, Tracy (2007). "Before the abyss: Agamben on Heidegger and the living." *Continental Philosophy Review*, *40*(1), 1–16.

Cornell, Heather, John M. Marzluff, and Shannon Pecoraro (2012). "Social learning spreads knowledge about dangerous humans among American crows." *Proc. Royal Soc. B*, *279*(1728), 499–508.

Coscarella, Mariano A., Gabriela Bellazzi, María Leoní Gaffet, Marisa Berzano, and Mariana Degrati (2015). "Technique used by killer whales (*Orcinus orca*) when hunting for dolphins in Patagonia, Argentina." *Aquatic Mammals*, *41*(2), 192–7.
Cossins, Daniel (2015). "The impersonators." *New Scientist*, *226*(3019), 41–3.
Crampton, Joel, Celine H. Frére, and Dominique A. Potvin (2022). "Australian Magpies *Gymnorhina tibicen* cooperate to remove tracking devices." *Australian Field Ornithology*, *39*, 7–11.
Crockford, Catherine, Roman M. Wittig, and Klaus Zuberbühler (2017). "Vocalizing in chimpanzees is influenced by social-cognitive processes." *Science Advances*, *3*(11), e1701742.
Cucherousset, Julien, Stéphanie Boulêtreau, Frédéric Azémar, Arthur Compin, Mathieu Guillaume, and Frédéric Santoul (2012). "'Freshwater killer whales': Beaching behavior of an alien fish to hunt land birds." *PLoS ONE*, *7*(12), e50840.
Cykowski, Beth (2021). *Heidegger's Metaphysical Abyss. Between the Human and the Animal*. Oxford University Press. Oxford.
DaDagenais, Paule, Sean Hensman, Valerie Haehler, and Michel C. Milinkovitch (2021). "Elephants evolved strategies reducing the biomechanical complexity of their trunk." *Current Biology*, *31*(21), 4727–37.
Dahlstrom, Daniel O. (2013). *The Heidegger Dictionary*. Bloomsbury. London & New York.
Danaher-Garcia, Nicole, Richard Connor, Gavin Fay, Kelly Melillo-Sweeting, and Kathleen M. Dudzinski (2022). "The partial merger of two dolphin societies." *Royal Society Open Sci.*, *9*(8), 211963.
Danchin, Etienne, Sabine Nöbel, Arnaud Pocheville, Anne-Cecile Dagaeff, Léa Demay, Mathilde Allphand, Sarah Ranty-Roby, Lara van Renssen, Magdalena Monier, Eva Gazagne, Mélanie Allain, and Guillaume Isabel (2018). "Cultural flies: Conformist social learning in fruitflies predicts long-lasting mate-choice traditions." *Science*, *362*(6418), 1025–30.
Darwin, Charles (2006). "On the origin of species." *In*: Edward O. Wilson (Ed.), *From So Simple a Beginning. The Four Great Books of Charles Darwin* (pp. 441–760). W.W. Norton. New York & London.
Davidson, Eric H. (1968). *Gene Activity in Early Development*. Academic Press. New York.
Davis, Kenneth L., and Christian Montag (2019). "Selected principles of Pankseppian affective neuroscience." *Frontiers in Neuroscience*, *12*, 1025.
de Kort, Selvino, Anthony Dickinson, and Nicola S. Clayton (2005). "Retrospective cognition by food-caching western scrub-jays." *Learning and Motivation*, *36*(2), 159–76.
de Waal, Frans (1990). *Peacemaking Among Primates*. Harvard University Press. Cambridge, MA, & London.
de Waal, Frans (1996). *Good Natured. The Origins of Right and Wrong in Humans and Other Animals*. Harvard University Press. Cambridge, MA, & London.
de Waal, Frans (2013). *The Bonobo and the Atheist*. W.W. Norton.
de Waal, Frans (2016a). *Are We Smart Enough to Know How Smart Animals Are?* W.W. Norton. New York & London.
de Waal, Frans (2016b, April 8). "What I learned from tickling apes." *New York Times*, Opinion.
de Waal, Frans (2019a). "Fish, mirrors, and a gradualist perspective on self-awareness." *PLoS Biology*, *17*(2), e3000112.
de Waal, Frans (2019b). *Mama's Last Hug*. W. W. Norton. New York & London.
de Waal, Frans (2022). *Different. Gender Through the Eyes of a Primatologist*. W.W. Norton. New York & London.

de Waal, Frans, Christine M. Korsgaard, Philip Kitcher, and Peter Singer (2006). *Primates and Philosophers. How Morality Evolved*. Princeton University Press. Princeton, NJ.

DeAngelis, Ross S., and Hans A. Hofman (2023). "The spread of fear in an empathetic fish. An evolutionarily ancient signaling pathway mediates emotional contagion." *Science*, *379*(6638), 1186–7.

Deloria Jr., Vine (1973). *God Is Red*. Dell Publishing. New York.

Demmerling, Christoph (2021). "Discourse (*Rede*)." *In*: Mark A. Wrathall (Ed.), *The Cambridge Heidegger Lexicon* (pp. 236–9). Cambridge University Press. Cambridge & London.

Derrida, Jacque (2008). *The Animal That Therefore I Am* (David Wills, Trans.) (Kindle ebook [www.amazon.com; downloaded November 23, 2015]). Fordham University Press. New York.

DeSalle, Rob, and Hans Bachor (2020). "Color without eyes." *Natural History Magazine*, *128*(6), 34–9.

DeSilva, Jeremy (2021). *First Steps. How Upright Walking Made Us Human*. HarperCollins. New York.

DeWeerdt, Sarah (2019, April 2). "Social networks a key to orca survival." *Encyclopedia of Puget Sound* (https://www.eopugetsound.org/magazine/is/orca-social-bonds). Tacoma, WA.

Dewey, John (1926, September). "The ethics of animal experimentation." *The Atlantic*.

Dong, Shihao, Tao Lin, James C. Nieh, and Ken Tan (2023). "Social signal of the waggle dance in honey bees." *Science*, *379*(6636), 1015–18.

Dotson, Nicholas M., and Michael M. Yartsev (2021). "Nonlocal spatiotemporal representation in the hippocampus of freely flying bats." *Science*, *373*(6551), 242–7.

Dreyfus, Hubert L. (1991). *Being-in-the-World. A Commentary on Heidegger's Being and Time*. MIT Press. Cambridge & London.

Edsinger, Eric, and Gül Dölen (2018). "A conserved role for serotonergic neurotransmission in mediating social behavior in octopus." *Current Biology*, *28*(19), 3136–42.

Elden, Stuart (2006). "Heidegger's Animals." *Continental Philosophy Review*, *39*(3), 273–91.

Elliser, Cindy R., and Denise L. Herzing (2016). "Long-term interspecies association patterns of Atlantic bottlenose dolphins, *Tursiops truncatus*, and Atlantic spotted dolphins, *Stenella frontalis*, in the Bahamas." *Marine Mammal Science*, *32*(1), 38–56.

Elwood, Richard W. (2012). "Evidence for pain in decapod crustaceans." *Animal Welfare*, *21*(s2), 23–7.

Elwood, Richard W. (2021). "Potential pain in fish and decapods: Similar experimental approaches and similar results." *Frontiers in Veterinary Science*, *8*, article 631151.

Engelland, Chad (2015). "Heidegger and the human difference." *Journal of the American Philosophical Association*, *1*(1), 175–93.

Festa-Bianchet, Marco (2018). "Learning to migrate." *Science*, *361*(6406), 972–3.

Finn, Julian K., Tom Tregenza, and Mark D. Norman (2009). "Defensive tool use in a coconut- carrying octopus." *Current Biology*, *19*(23), R1070.

Flores, Dan (2016). *Coyote America. A Natural and Supernatural History*. Basic Books. New York.

Foley, Nicole M., Victor C. Mason, Andrew J. Harris, Kevin R. Bredemeyer, Joana Damas, Harris A. Lewin, Eduardo Eizrik, John Gatesy, Elinor K. Karlsson, Kerstin Lindblad-Toh, Zoonomia Consortium, Mark S. Springer, and William J. Murphy (2023). "A genomic timescale for placental mammal evolution." *Science*, *380*(6643), eabl8189.

Foltz, Bruce V. (1993). "Heidegger, ethics, and animals." *Between the Species*, *9*(2), article 8, 84–9.

Foltz, Bruce V. (1995). *Inhabiting the Earth. Heidegger, Environmental Ethics, and the Metaphysics of Nature* (Kindle ebook [www.amazon.com; downloaded June 26, 2017]). Humanity Books. Amherst, NY.

Foltz, Bruce V. (2014). *The Noetics of Nature. Environmental Philosophy and the Holy Beauty of the Visible*. Fordham University Press. New York.
Foster, Craig (2020). *My Octopus Teacher*. Netflix documentary film. DVD.
Foster, Craig (2024, April 23). "An octopus took my camera, and the images changed the way I see the world." *New York Times*, Opinion / Guest Essay.
Fragaszy, Dorothy Munkenbeck, Nalina Aiempichitkijkarn, Yonat Eshchar, Madhur Mangalam, Patrícia Izar, Briseida Resende, and Elisabetta Visalberghi (2023). "The development of expertise at cracking palm nuts by wild bearded capuchin monkeys, *Sapajus libidinosus*." *Animal Behaviour, 197*(1), 1–14.
Fraser, Orlaith N., and Thomas Bugnyar (2010). "Do ravens show consolation? Responses to distressed others." *PLoS ONE, 5*(5), e10605.
Fried, Gregory (2017). "Whitewashed with moralism: On Heidegger's anti-Americanism and anti-Semitism." *In*: Elad Lapidot, Micha Brumli, and Elan Reisner (Eds.), *Heidegger and Jewish Thought. Difficult Others* (p. 313). Rowman and Littlefield International. London.
Fugazza, Claudia, Àkos Pogány, and Ádám Miklósi (2016). "Recall of others' actions after incidental encoding reveals episodic-like memory in dogs." *Current Biology, 26*(23), 3209–13.
Gadamer, Hans-Georg (1994). *Heidegger's Ways* (John W. Stanley, Trans.). State University of New York. Albany, NY.
Galpayage Dona, Hirune Samadi, Cwyn Solvi, Amelia Kowalewska, Kaarle Mäkelä, HaDi MaBouDi, and Lars Chittka (2022). "Do bumble bees play?" *Animal Behaviour, 194*, 239–51.
Garrett, Aaron (2011). "Animals and ethics in the history of modern philosophy." *In*: Tom L. Beauchamp, and R.G. Fry (Eds.), *The Oxford Handbook of Animal Ethics* (pp. 61–87). Oxford University Press. New York.
Godfrey-Smith, Peter (2016). *Other Minds. The Octopus, the Sea, and the Deep Origins of Consciousness* (Kindle ebook [www.amazon.com; downloaded December 8, 2016]). Farrar, Straus and Giroux.
Godfrey-Smith, Peter (2020). *Metazoa. Animal Life and the Birth of the Mind*. Farrar, Straus and Giroux. New York.
Gould, Stephen Jay (1977). *Ontogeny and Phylogeny*. Harvard University Press. Cambridge, MA, & London.
Gould, Stephen Jay (2002). *The Structure of Evolutionary Theory*. Harvard University Press. Cambridge, MA, & London.
Gould, Stephen Jay, and Elisabeth A. Lloyd (1999). "Individuality and adaptation across levels of selection: How shall we name and generalize the unit of Darwinism?" *Proc. Natl. Acad. Sci. U.S.A., 96*(21), 11904–9.
Graham, Kirsty E., and Catherine Hobaiter (2023). "Towards a great ape dictionary: Inexperienced humans understand common nonhuman ape gestures." *PLoS Biology, 21*(1), e3001939.
Grandi, M. Florencia, Rocío Loizaga de Castro, and Enrique A. Crespo (2012). "Killer whales attack on South American sea lion associated with a fishing vessel: Predator and prey tactics." *Lat. Am. J. Aquat. Res., 40*(4), 1072–6.
Griffin, Donald R. (1981). *The Question of Animal Awareness. Evolutionary Continuity of Mental Experience*. William Kaufmann. Los Altos, CA.
Griffiths, Siân W., and Anne E. Magurran (1997). "Familiarity in schooling fish: How long does it take to acquire?" *Animal Behaviour, 53*(5), 945–9.
Grimm, David (2019). "Ready to pounce." *Science, 364*(6440), 522–5.

Grosenick, Logan, Tricia S. Clement, and Russell Fernald (2007). "Fish can infer social rank by observation alone." *Nature, 445*(7126), 429–32.
Gunasekaram, Cassandra, Federico Battiston, Onkar Sadekar, Cecilia Padilla-Iglesias, Maria A. van Noordwijk, Reinhard Furrer, Andrea Manica, Jaume Bertranpetit, Andrew Whiten, Carel P. van Schaik, Lucio Vinicius, and Andrea Bamberg Migliano (2024). "Population connectivity shapes the distribution and complexity of chimpanzee cumulative culture." *Science, 386*(6724), 920–5.
Habermas, Jürgen (1977). "Martin Heidegger: On the publication of lectures from the year 1935." *Graduate Faculty Philosophy Journal, 6*(2), 155–80.
Habl, Cornelia, and Alice Marie Isabel Auersperg (2017). "The keybox: Shape-frame fitting during tool use in Goffin's cockatoos (*Cacatua goffiniana*)." *PLoS ONE, 12*(11), e0186859.
Haeckel, Ernst (1905). *The Riddle of the Universe at the Close of the Nineteenth Century* (Joseph McCabe, Trans.). Harper and Brothers. New York & London.
Harjo, Joy, LeAnne Howe, and Jennifer Elise Foerster (Eds.). (2020). *When the Light of the World was Subdued, Our Songs Came Through*. W.W. Norton. New York.
Harten, Lee, Amitay Katz, Aya Goldshtein, Michal Handel, and Yossi Yovel (2020). "The ontogeny of a mammalian cognitive map in the real world." *Science, 369*(6500), 194–7.
Hatab, Lawrence J. (2014). "Dasein, the early years: Heideggerian reflections on childhood." *International Philosophical Quarterly, 54*(4), 379–91.
Hatzopoulos, Antonis K., and Jerome C. Regier (1987). "Evolutionary changes in the developmental expression of silkmoth chorion genes and their morphological consequences." *Proc. Natl. Acad. Sci. U.S.A., 84*(2), 479–83.
Haugeland, John (2013). *Dasein Disclosed. John Haugeland's Heidegger*. Harvard University Press. Cambridge, MA, & London.
Heidegger, Martin (1962). *Being and Time* (J. Macquarrie, and E. Robinson, Trans.). Harper and Row. New York.
Heidegger, Martin (1971a). "Language in the poem. A discussion on Georg Trakl's poetic work" (Peter D. Hertz, Trans.). *In*: *On the Way to Language* (pp. 159–98). HarperOne. New York.
Heidegger, Martin (1971b). "The nature of language" (Peter D. Hertz, Trans.). *In*: *On the Way to Language* (pp. 111–36). HarperOne. New York.
Heidegger, Martin (1972a). *On Time and Being* (Joan Stambaugh, Trans.). Harper and Row. New York.
Heidegger, Martin (1972b). "Time and being" (Joan Stambaugh, Trans.). *In*: *On Time and Being* (pp. 1–24). Harper and Row. New York.
Heidegger, Martin (1976). *What Is Called Thinking?* (J. Glenn Gray, Trans.). HarperCollins. New York.
Heidegger, Martin (1985). *Schelling's Treatise on the Essence of Human Freedom* (Joan Stambaugh, Trans.). Ohio University Press. Athen, OH.
Heidegger, Martin (1988). *The Basic Problems of Phenomenology* (Albert Hofstadter, Trans.). Indiana University Press. Bloomington & Indianapolis, IN.
Heidegger, Martin (1992a). *Prolegomena to the History of the Concept of Time* (Theodore Kisiel, Trans.). Indiana University Press. Bloomington & Indianapolis, IN.
Heidegger, Martin (1992b). *Parmenides* (André Schuwer, and Richard Rojcewicz, Trans.). Indiana University Press. Bloomington & Indianapolis, IN.
Heidegger, Martin (1994). *Basic Questions of Philosophy. Selected "Problems" of "Logic"* (Richard Rojcewicz, and André Schuwer, Trans.). Indiana University Press. Bloomington & Indianapolis, IN.

Heidegger, Martin (1995). *The Fundamental Concepts of Metaphysics. World, Finitude, Solitude* (William McNeill, and Nicholas Walker, Trans.). Indiana University Press. Bloomington & Indianapolis, IN.

Heidegger, Martin (1998a). "Introduction to 'what is metaphysics?'" (Walter Kaufmann, Trans.). *In*: William McNeill (Ed.), *Pathmarks* (pp. 277–90). Cambridge University Press. Cambridge.

Heidegger, Martin (1998b). "Letter on 'humanism'" (Frank A. Capuzzi, Trans.). *In*: William McNeill (Ed.), *Pathmarks* (pp. 239–76). Cambridge University Press. Cambridge.

Heidegger, Martin (1998c). "On the essence of ground" (William McNeill, Trans.). *In*: William McNeill (Ed.), *Pathmarks* (pp. 97–135). Cambridge University Press. Cambridge.

Heidegger, Martin (2001). *Zollikon Seminars. Protocols–Conversations–Letters* (Frans Mayr, and Richard Askay, Trans.). Northwestern University Press. Evanston, IL.

Heidegger, Martin (2002a). "The age of the world picture" (+ Appendices) (Julian Young, and Kenneth Haynes, Trans.). *In*: Julian Young, and Kenneth Haynes (Eds.), *Off the Beaten Track* (pp. 57–85). Cambridge University Press. Cambridge.

Heidegger, Martin (2002b). "The origin of the work of art" (Julian Young, and Kenneth Haynes, Trans.). *In*: Julian Young, and Kenneth Haynes (Eds.), *Off the Beaten Track* (pp. 1–56). Cambridge University Press. Cambridge.

Heidegger, Martin (2002c). "Wilhelm Dilthey's research and the struggle for a historical worldview (1925)" (Charles Bambach, John Protevi, Roderick M. Stewart, Harry S. Taylor, Hans W. Uffelmann, and John van Buren, Trans.). *In*: John van Buren (Ed.), *Supplements. From the Earliest Essays to Being and Time and Beyond* (pp. 147–76). State University of New York Press. Albany, NY.

Heidegger, Martin (2003a). "Overcoming metaphysics" (Joan Stambaugh, Trans.). *In*: *The End of Philosophy* (pp. 84–110). University of Chicago Press. Chicago, IL.

Heidegger, Martin (2003b). "Seminar in Le Thor 1969" (Andrew Mitchell, and François Raffoul, Trans.). *In*: *Four Seminars. Le Thor 1966, 1968, 1969, Zähringen 1973* (pp. 35–63). Indiana University Press. Bloomington & Indianapolis, IL.

Heidegger, Martin (2007). "On the essence of truth (Pentecost Monday, 1926)" (Theodore Kisiel, Trans.). *In*: Theodore Kisiel, Thomas (Ed.), *Becoming Heidegger. On the Trail of His Early Occasional Writings, 1910–1927* (pp. 275–87). Northwestern University Press. Evanston, IL.

Heidegger, Martin (2008). "What calls for thinking?" (Albert Hofstadter, Trans.). *In*: *Basic Writings* (pp. 365–91). Harper Perennial. New York.

Heidegger, Martin (2009). *Basic Concepts of Aristotelian Philosophy* (Robert D. Metcalf, and Mark B. Tanzer, Trans.). Indiana University Press. Bloomington & Indianapolis, IL.

Heidegger, Martin (2010). *The Phenomenology of Religious Life* (Matthias Fritsch, and Jennifer Anna Gosetti-Ferencei, Trans.). Indiana University Press. Bloomington & Indianapolis, IL.

Heidegger, Martin (2012a). "Basic principles of thinking: Freiburg lectures 1957" (Andrew J. Mitchell, Trans.). *In*: John Sallis (Ed.), *Bremen and Freiburg Lectures* (pp. 75–166). Indiana University Press. Bloomington & Indianapolis, IL.

Heidegger, Martin (2012b). *Contributions to Philosophy (Of the Event)* (Richard Rojcewicz, and Daniela Vallega-Neu, Trans.). Indiana University Press. Bloomington & Indianapolis, IL.

Heidegger, Martin (2013a). *Basic Problems of Phenomenology (Winter Semester 1919 / 1920)* (Scott M. Campbell, Trans.). Bloomsbury Academic. London.

Heidegger, Martin (2013b). "Building dwelling thinking" (Albert Hofstadter, Trans.). *In*: *Poetry, Language, Thought* (pp. 141–59). HarperCollins Publishers. New York.

Heidegger, Martin (2013c). *The Essence of Truth. On Plato's Cave Allegory and Theaetetus* (Ted Sadler, Trans.). Bloomsbury Academic. London.

Heidegger, Martin (2013d). *The Event* (Richard Rojcewicz, Trans.). Indiana University Press. Bloomington & Indianapolis, IN.

Heidegger, Martin (2013e). *The Question Concerning Technology and Other Essays* (William Lovitt, Trans.). Harper Perennial Modern Thought. New York.

Heidegger, Martin (2013f). "The Thing" (Albert Hofstadter, Trans.). *In*: *Poetry, Language, Thought* (pp. 161–84). Harper Perennial Modern Thought. New York.

Heidegger, Martin (2014). *Introduction to Metaphysics* (Gregory Fried, and Richard Polt, Trans.). Yale University Press. New Haven, CT, & London.

Heidegger, Martin (2015). *The History of Beyng* (William McNeill, and Jeffrey Powell, Trans.). Indiana University Press. Bloomington & Indianapolis, IN.

Heidegger, Martin (2016a). "*Agchibasie*: A triadic conversation on a country path between a Scientist, a Scholar, and a Guide" (Bret W. Davis, Trans.). *In*: *Country Path Conversations* (pp. 1–104). Indiana University Press. Bloomington & Indianapolis, IN.

Heidegger, Martin (2016b). *Interpretation of Nietzsche's Second Untimely Meditation* (Ullrich Haase and Mark Sinclair, Trans.). Indiana University Press. Bloomington & Indianapolis, IN.

Heidegger, Martin (2016c). *Mindfulness* (Parvis Emad, and Thomas Kalary, Trans.). Bloomsbury Academic. London.

Heidegger, Martin (2017). *Ponderings VII-XI. Black Notebooks 1938–1939.* (Richard Rojcewicz, Trans.). Indiana University Press. Bloomington & Indianapolis, IN.

Heidegger, Martin (2024). *Introduction to Philosophy* (William McNeill, Trans.). Indiana University Press. Bloomington & Indianapolis, IN.

Heidegger, Martin, and Eugen Fink (1993). *Heraclitus Seminar* (Charles H. Seibert, Trans.). Northwestern University Press. Evanston, IL.

Heinrich, Bernrd (2016). *One Wild Bird at a Time. Portraits of Individual Lives* (Kindle ebook [www.amazon.com; downloaded October 20, 2016]). Houghton Mifflin Harcourt. Boston, MA, & New York.

Heinrich, Bernrd (2019). "Bridging the divide. Foiled by spoiled meat." *Natural History Magazine*, *127*(6), 9–11.

Heinrich, Bernrd (2024). "A peeved pileated. Reflections of an angry bird." *Natural History Magazine*, *132*(7), 12–13.

Hennig, Willi (1979). *Phylogenetic Systematics* (D.D. Davis, and R. Zangerl, Trans.). University of Illinois Press. Urbana & Chicago, IL.

Hillemacher, Sonja, Sebastian Ocklenburg, Onur Güntürkün, and Inga Tiemann (2023). "Roosters do not warn the bird in the mirror: The cognitive ecology of mirror self-recognition." *PLoS ONE*, *18*(10), e1290416.

Hölldobler, Bert, and Edward O. Wilson (2009). *The Superorganism. The Beauty, Elegance, and Strangeness of Insect Societies.* W. W. Norton. New York.

Holló, Gábor, and Mihály Novák (2012). "The manoeuvrability hypothesis to explain the maintenance of bilateral symmetry in animal evolution." *Biol Direct*, *7*(1), 22.

Holzhaider, Jennifer C., M.D. Sibley, A.H. Taylor, P.J. Singh, R.D. Gray, and G.R. Hunt (2011). "The social structure of New Caledonian crows." *Animal Behaviour*, *81*(1), 83–92.

Honegger, Kyle, and Benjamin de Bivort (2018). "Stochasticity, individuality and behavior." *Current Biology*, *28*(1), R8–R12.

Hough, Alexander R., John Case, and Jean Geary Boal (2016). "Learned control of body patterning in cuttlefish *Sepia officinalis* (Cephalopoda)." *Journal of Molluscan Studies*, *82*(3), 427–31.

Howard, Scarlett R., Aurore Avarguès-Weber, Jair E. Garcia, Andrew D. Greentree, and Adrian G. Dyer (2018). "Numerical ordering of zero in honey bees." *Science*, *360*(6393), 1124–6.

Howard, Scarlett, Aurore Avarguès-Weber, Jair E. Garcia, Andrew D. Greentree, and Adrian G. Dyer (2019). "Numerical cognition in honeybees enables addition and subtraction." *Science Advances*, *5*(2), eaav0961.

Hunt, Gavin R. (1996). "Manufacture and use of hook-tools by New Caledonian crows." *Nature*, *379*(6562), 249–51.

Hunt, Gavin R., and Russel D. Gray (2002). "Species-wide manufacture of stick-type tools by New Caledonian Crows." *Emu*, *102*(4), 349–53.

Iglesias, Teresa L., Jean G. Boal, Marcos G. Frank, Jochen Zeil, and Roger T. Hanlon (2019). "Cyclic nature of the REM sleep-like state in cuttlefish *Sepia officinalis*." *Journal of Experimental Biology*, *222*(1), jeb174862.

Jacobs, Ivo, and Mathias Osvath (2023). "Tool use and tooling in ravens (*Corvus corax*): A review and novel observations." *Ethology*, *129*(3), 169–81.

Jagiello, Zuzanna, Łukasz Dylewski, and Marta Szulkin (2024). "The plastic homes of hermit crabs in the Anthropocene." *Science of the Total Environment*, *913*(11), 168959.

Jesmer, Brett R., Jerod A. Merkle, Jacob R. Goheen, Ellen O. Aikens, Jeffrey L. Beck, Alyson B. Courtemanch, Mark A. Hurley, Douglas E. McWhirter, Hollie M. Miyasaki, Kevin L. Monteith, and Matthew J. Kauffman (2018). "Is ungulate migration culturally transmitted? Evidence of social learning from translocated animals." *Science*, *361*(6406), 1023–5.

Jones, Bethany, Elizabeth Bukoski, Lynn Nadel, and Jean-Marc Fellous (2012). "Remaking memories: Reconsolidation updates positively motivated spatial memory in rats." *Learning and Memory*, *19*(3), 91–8.

Kaas, Jon H. (2009). "Evolution of the brain in mammals." *In*: M.D. Binder, N. Hirokawa, and U. Windhorst (Eds.), *Encyclopedia of Neuroscience* (pp. 1292–5). Springer. Berlin & Heidelberg.

Kabadayi, C., and M. Osvath (2017). "Ravens parallel great apes in flexible planning for tool-use and bartering." *Science*, *357*(6347), 202–4.

Kano, Fumihiro, and Satoshi Hirata (2015). "Great apes make anticipatory looks based on long-term memory of single events." *Current Biology*, *25*(19), 2513–17.

Karihaloo, B.L., K. Zhang, and J. Wang (2013). "Honeybee combs: How the circular cells transform into rounded hexagons." *Journal of the Royal Society Interface*, *10*(86), 20130299.

Kays, Roland, and Robert Michelson (2018). "Coyote continent." *Natural History Magazine*, *126*(July/August), 22–7.

Kendal, Jeremy R., Luke Rendell, Thomas W. Pike, and Kevin N. Laland (2009). "Nine-spined sticklebacks deploy a hill-climbing social learning strategy." *Behavioral Ecology*, *20*(2), 238–44.

Kenward, Ben, Christian Schloegl, Christian Rutz, Alexander A.S. Weir, Thomas Bugnyar, and Alex Kacelnik (2011). "On the evolutionary and ontogenetic origins of tool oriented behaviour in New Caledonian crows (*Corvus moneduloides*)." *Biological Journal of the Linnean Society*, *102*(4), 870–7.

Kimmerer, Robin Wall (2013). *Braiding Sweetgrass. Indigenous Wisdom, Scientific Knowledge, and the Teachings of Plants*. Milkweed Editions. Minneapolis, MN.

Kimmerer, Robin Wall (2024). *The Serviceberry. Abundance and Reciprocity in the Natural World.* Scribner. New York.
King, Barbara J. (2019a). "Deception in the wild." *Scientific American, 321*(3), 50–4.
King, Barbara J. (2019b). "The orca's sorrow." *Scientific American, 320*(3), 30–5.
Kisiel, Theodore (2001). "Heidegger's Philosophical Geopolitics in the Third Reich." *In*: Richard Polt, and Gregory Fried (Eds.), *A Companion to Heidegger's Introduction to Metaphysics* (pp. 226–49). Yale University Press. New Haven, CT, & London.
Klump, Barbara C., James J.H. St Clair, and Christian Rutz (2021). "New Caledonian crows keep 'valuable' hooked tools safer than basic non-hooked tools." *eLife, 10*, e64829.
Klump, Barbara C., Shoko Sugasawa, James J.H. St Clair, and Christian Rutz (2015). "Hook tool manufacture in New Caledonian crows: Behavioural variation and the influence of raw materials." *BMC Biology, 13*(1), 97.
Klump, Barbara C., John M. Martin, Sonja Wild, Jana K. Hörsch, Richard E. Major, and Lucy M. Aplin (2021). "Innovation and geographic spread of a complex foraging culture in an urban parrot." Science, *373*(6553), 456–60.
Kohda, Masanori, Takashi Hotta, Tomohiro Takeyama, Satoshi Awata, Hirokazu Tanaka, Jun-Ya Asai, and Alex L. Jordan (2019). "If a fish can pass the mark test, what are the implications for consciousness and self- awareness testing in animals?" *PLoS Biology, 17*(2), e3000021.
Kohda, Masanori, Redouan Bshary, Naoki Kubo, Satoshi Awata, Will Sowersby, Kento Kawasaka, Taiga Kobayashi, and Shumpei Sogawa (2023). "Cleaner fish recognize self in a mirror via self-face recognition like humans." *Proc. Natl. Acad. Sci. U.S.A., 120*(7), e2208420120.
Kohda, Masanori, Shumpei Sogawa, Alex L. Jordan, Naoki Kubo, Satoshi Awata, Shun Satoh, Taiga Kobayashi, Akane Fujita, and Redouan Bshary (2022). "Further evidence for the capacity of mirror self-recognition in cleaner fish and the significance of ecologically relevant marks." *PLoS Biology, 20*(2), e3001529.
Korsgaard, Christine M. (2018). *Fellow Creatures. Our Obligations to the Other Animals.* Oxford University Press. New York.
Krauss, Lawrence M. (2012). *A Universe from Nothing. Why There Is Something Rather than Nothing.* Free Press. New York.
Krupenye, Christopher, Fumihiro Kano, Satoshi Hirata, Josep Call, and Michael Tomasello (2016). "Great apes anticipate that other individuals will act according to false beliefs." *Science, 354*(6308), 110–14.
Kuo, Tzu-Hsin, and Chuan-Chin Chiao (2020). "Learned valuation during forage decision-making in cuttlefish." *Royal Soc. Open Sci., 7*(12), 201602.
Langergraber, Kevin E., David P. Watts, Linda Vigilant, and John C. Mitani (2017). "Group augmentation, collective action, and territorial boundary patrols by male chimpanzees." *Proc. Natl. Acad. Sci. U.S.A., 114*(28), 7337–42.
Laumer, I.B., S.L. Winkler, F. Rossano, and E.A. Cartmill (2024). "Spontaneous playful teasing in four great ape species." *Proc. Royal Soc. B, 291*(2016), 20232345.
Law, Yao-Hua (2021). "Long live the queen." *Science, 371*(6536), 1302–5.
Lee, Won Young, Sang-im Lee, Jae Chun Choe, and Piotr G. Jablonski (2011). "Wild birds recognize individual humans: Experiments on magpies, *Pica pica*." *Animal Cognition, 14*(6), 817–15.
Lee, Won Young, Yeong-Deok Han, Sang-im Lee, Piotr G. Jablonski, Jin-Woo Jung, and Jeong-Hoon Kim (2016). "Antarctic skuas recognize individual humans." *Animal Cognition, 19*(4), 861–5.

Lethmate, Jügen (1982). "Tool-using skills of orang-utans." *Journal of Human Evolution*, *11*(1), 49–50, IN5–IN8, 51–64.
Lewis, Laura S., Erin G. Wessling, Fumihiro Kano, Jeroen M.G. Stevens, Josep Cal, and Christopher Krupenye (2023). "Bonobos and chimpanzees remember familiar conspecifics for decades." *Proc. Natl. Acad. Sci. U.S.A.*, *120*(52), e2304903120.
Lewontin, Richard C. (2000). *The Triple Helix*. Harvard University Press. Cambridge, MA, & London.
Lewontin, Richard C. (2001). "Gene, organism and environment." *In*: Susan Oyama, Paul E. Griffiths, and Russell D. Gray (Eds.), *Cycles of Contingency. Developmental Systems and Evolution* (pp. 59–66). Bradford Books / MIT Press. Cambridge, MA, & London.
Lilly, G.R. (2008). "The decline, recovery, and collapse of Atlantic cod (*Gadus morhua*) off Labrador and eastern Newfoundland." *In*: G.H. Kruse, K. Drinkwater, J.N. Ianelli, J.S. Link, D.L. Stram, V. Wespestad, and D. Woodby (Eds.), *Resiliency of Gadid Stocks to Fishing and Climate Change* (pp. 67–87). Alaska Sea Grant, University of Alaska. Fairbanks, AK.
Losos, Jonathan B. (2023). *The Cat's Meow. How Cats Evolved from the Savanna to Your Sofa*. Viking. New York.
Loukola, Olli J., Anna Antinoja, Kaarle Mäkelä, Janette Arppi, Fei Peng, and Cwyn Solvi (2024). "Evidence for socially influenced and potentially actively coordinated cooperation by bumblebees." *Proc. Royal Soc. B*, *291*(2022), 20240055.
Madsen, Peter T., Ursula Siebert, and Coen P.H. Elemans (2023). "Toothed whales use distinct vocal registers for echolocation and communication." *Science*, *379*(6635), 928–33.
Maeterlinck, Maurice (1901). *The Life of the Bee*. Blue Ribbon Books. New York.
Magee, Barry, and Robert W. Elwood (2013). "Shock avoidance by discrimination learning in the shore crab (*Carcinus maenas*) is consistent with a criterion of pain." *Journal of Experimental Biology*, *216*(3), 353–8.
Magee, Barry, and Robert W. Elwood (2016). "No discrimination shock avoidance with sequential presentation of stimuli but shore crabs still reduce shock exposure." *Biology Open*, *5*(7), 883–8.
Mann, Janet, Brooke L. Sargeant, Jana J. Watson-Capps, Quincy A. Gibson, Michael R. Heithaus, Richard C. Connor, and Eric Patterson (2008). "Why do dolphins carry sponges?" *PLoS ONE*, *3*(12), e3868.
Martinho III, Antone, and Alex Kacelnik (2016). "Ducklings imprint on the relational concept of 'same or different.'" *Science*, *353*(6296), 286.
Marzluff, John M., Jeff Walls, Heather N. Cornell, John C. Withey, and David P. Craig (2010). "Lasting recognition of threatening people by wild American crows." *Animal Behaviour*, *79*(3), 699–707.
Masud, M.H., M. La Mantia, and P. Dabnichki (2022). "Estimate of Strouhal and Reynolds numbers for swimming penguins." *Journal of Avian Biology*, *2022*(2), e02886.
Mather, Jennifer A. (2008). "Cephalopod consciousness: Behavioural evidence." *Consciousness and Cognition*, *17*(1), 37–48.
Mayr, Ernst (1982). *The Growth of Biological Thought. Diversity, Evolution, and Inheritance*. Belknap Press. Cambridge & London.
McAuliffe, Katherine, Lindsey A. Drayton, Amanda Royka, Mélisande Aellen, Laurie R. Santos, and Redouan Bshary (2021). "Cleaner fish are sensitive to what their partners can and cannot see." *Communications Biology*, *4*(1), 1127.

McCoy, Dakota E., Martina Schiestl, Patrick Neilands, Rebecca Hassall, Russel D. Gray, and Alex H. Taylor (2019). "New Caledonian crows behave optimistically after using tools." *Current Biology, 29*(16), 2737–42.
McGaugh, James L. (2013). "Making last memories: Remembering the significant." *Proc. Natl. Acad. Sci. U.S.A., 110, suppl. 2*, 10402–7.
McGhee, George R. (2021). "Convergent evolution of agriculture in bilaterian animals: An adaptive landscape perspective." *In*: Ted R. Schultz, Richard Gawne, and Peter N. Peregrine (Eds.), *The Convergent Evolution of Agriculture in Humans and Insects* (pp. 7–30). The MIT Press. Cambridge, MA, & London.
McNeill, William (1999). "Life beyond the organism: Animal being in Heidegger's Freiburg Lectures, 1929–1930." *In*: H. Peter Steeves (Ed.), *Animal Others. On Ethics, Ontology, and Animal Life* (pp. 197–248). State University of New York Press. Albany, NY.
Menz, Myles H.M., Martina Scacco, Hans-Martin Bürki-Spycher, Hannah J. Williams, Don R. Reynolds, Jason W. Chapman, and Martin Wikelski (2022). "Individual tracking reveals long-distance flight-path control in a nocturnally migrating moth." *Science, 377*(6607), 764–8.
Messina, Andre, Davide Potrich, Ilaria Schiona, Valeria Anna Sovrano, and Giorgio Vallortigara (2021). "The sense of number in fish, with particular reference to its neurobiological bases." *Animals, 11*(11), 3072.
Millsopp, Sarah, and Peter Laming (2008). "Trade-offs between feeding and shock avoidance in goldfish (Carassius auratus)." *Applied Animal Behaviour Science, 113*(1–3), 247–54.
Mindell, David P. (2024). "Coevolution of humans and technology. We have always been technologists." *Natural History Magazine, 132*(7), 22–7.
Mitchell, Andrew J. (2011). "Heidegger's later thinking of animality: The end of world poverty." *Gatherings: The Heidegger Circle Annual, 1*, 74–85.
Mitchell, Andrew J. (2015). *The Fourfold. Reading the Late Heidegger*. Northwestern University Press. Evanston, IL.
Monsó, Susana (2024). *Playing Possum. How Animals Understand Death*. Princeton University Press. Princeton & Oxford.
Montague, Michael J., Gang Li, Barbara Gandolfi, Razib Khan, Bronwen L. Aken, Steven M.J. Searle, Patrick Minx, LaDeana W. Hillier, Daniel C. Koboldt, Brian W. Davis, Carlos A. Driscoll, Christina S. Barr, Kevin Blackistone, Javier Quilez, Belen Lonente-Galdos, Tomas Marques-Bonet, Can Alkan, Gregg W.C. Thomas, Matthew W. Hahn, Marilyn Menotti-Raymond, Stephen J. O'Brien, Richard K. Wilson, Leslie A. Lyons, William J. Murphy, and Wesley C. Warren (2014). "Comparative analysis of the domestic cat genome reveals genetic signatures underlying feline biology and domestication." *Proc. Natl. Acad. Sci. U.S.A., 111*(48), 17230–5.
Morrison, Rachel, and Diana Reiss (2018). "Precocious development of self-awareness in dolphins." *PLoS One, 13*(1), e0189813.
Morse, Peter, and Christine L. Huffard (2022). "Chemotactile social recognition in the blue-ringed octopus, *Hapalochlaena maculosa*." *Marine Biology, 169*(8), 99.
Murphy, Heather (2017, October 16). "Fish depression is not a joke." *New York Times*, Science / Trilobites.
Narayan, R.K., and M. Verma (2019). "Encephalization." *In*: J. Vonk, and T. Shackelford (Eds.), *Encyclopedia of Animal Cognition and Behavior* (pp. 1–3). Springer.
Nesher, Nir, Guy Levy, Frank W. Grasso, and Binyamin Hochner (2014). "Self-recognition mechanism between skin and suckers prevents octopus arms from interfering with each other." *Current Biology, 24*(11), 1271–5.

Newport, Cait, Guy Wallis, Yarema Reshitnyk, and Ulrike E. Siebeck (2016). "Discrimination of human faces by archerfish *(Toxotes chatareus).*" *Scientific Reports, 6*(1), 27523.

Nicol, Christine J. (2015). *The Behavioural Biology of Chickens*. CABI. Wallingford & Boston, MA.

Nietzsche, Friedrich Wilhelm (1997). "On the uses and disadvantages of history for life." *In*: Daniel Breazeale (Ed.), *Untimely Meditations* (pp. 57–124). Cambridge University Press. Cambridge.

Nowak, Martin A., and Roger Highfield (2011). *SuperCooperators. Altruism, Evolution, and Why We Need Each Other to Succeed.* Free Press. New York.

Nussbaum, Martha (2023). *Justice for Animals. Our Collective Responsibility.* Simon and Schuster. New York.

O'Donoghue, Ellen M., Matthew B. Broschard, John H. Freeman, and Edward A. Wasserman (2022). "The Lords of the Rings: People and pigeons take different paths mastering the concentric-rings categorization task." *Cognition, 218*, 104920.

O'Mara, M. Teague, and Christian Ziegler (2024). "Mammalian flight. The anatomy and evolution of bats." *Natural History Magazine, 132*(March), 16–23.

Okrent, Mark (2007). *Rational Animals. The Teleological Roots of Intentionality.* Ohio University Press. Athens, OH.

Olkowicz, Seweryn, Martin Kocourek, Radek K. Lučan, Michal Porteš, W. Tecumseh Fitch, Suzana Herculano-Houzel, and Pavel Němec (2016). "Birds have primate-like numbers of neurons in the forebrain." *Proc. Natl. Acad. Sci. U.S.A., 113*(26), 7255–60.

Panksepp, Jaak (2011a). "The basic emotional circuits of mammalian brains: Do animals have affective lives?" *Neuroscience and Biobehavioral Reviews, 35*(9), 1791–804.

Panksepp, Jaak (2011b). "Cross-Species affective neuroscience decoding of the primal affective experiences of humans and related animals." *PLoS ONE, 6*(9), e21236.

Panksepp, Jaak, and Lucy Biven (2012). *The Archaeology of Mind. Neuroevolutionary Origins of Human Emotions.* W.W. Norton. New York & London.

Pathak, Ayush, Steve Kett, and Massimiliano Marvasi (2019). "Resisting antimicrobial resistance: Lessons from fungus farming ants." *Trends in Ecology and Evolution, 34*(11), 974–6.

Pennisi, Elizabeth (2016). "The power of personality." *Science, 352*(6286), 644–7.

Pepperberg, Irene M. (2002). *The Alex Studies. Cognitive and Communicate Abilities of Grey Parrots.* Harvard University Press. Cambridge, MA, & London.

Pepperberg, Irene M. (2009). *Alex and Me.* Harper. New York.

Pepperberg, Irene M. (2020). "The comparative psychology of intelligence: Some thirty years later." *Frontiers in Psychology, 11*, 973.

Pepperberg, Irene M., and Virginia A. Rosenberger (2022). "Delayed gratification: A grey parrot (*Psittacus erithacus*) will wait for more tokens." *Journal of Comparative Psychology, 136*(1), 79–89.

Peter, Isabelle S., and Eric H. Davidson (2015). *Genomic Control Processes. Development and Evolution.* Academic Press. London.

Pineda Saldaña, César Alberto (2025). "The stone, the animal, and the human being rethought by Heidegger in the Black Notebooks." *Journal of Philosophy. Ibero-American University, 57*(158), 10–36.

Pitman, Robert L., Volker B. Deecke, Christine M. Gabriele, Mridula Srinivasan, Nancy Black, Judith Denkinger, John Durban, Elizabeth Mathews, Dena Matkin, Janet Neilson, Alisa Schulman-Janiger, Debra Shearwater, Peggy Stap, and Richard Ternullo (2017). "Humpback whales interfering when mammal-eating killer whales attack other species: Mobbing behavior and interspecific altruism?" *Marine Mammal Science*, *33*(1), 7–58.

Pokharel, Sanjeeta Sharma, Nachiketha Sharma, and Raman Sukumar (2022). "Viewing the rare through public lenses: Insights into dead calf carrying and other thanatological responses in Asian elephants using YouTube videos." *Royal Soc. Open Sci.*, *9*(5), 211740.

Polt, Richard (2006). *The Emergency of Being. On Heidegger's Contributions to Philosophy*. Cornell University Press. Ithaca, NY, & London.

Preston, Elizabeth (2022, May 17). "Elephants in mourning spotted on YouTube by scientists." *New York Times*, Trilobites.

Price, Hew (2013). *Expressivism, Pragmatism and Representationalism*. Cambridge University Press. Cambridge & London.

Prum, Richard O. (2017). "Beauty happens." *Natural History Magazine*, *125*(4), 24–9.

Prum, Richard O. (2018). *The Evolution of Beauty. How Darwin's Forgotten Theory of Mate Choice Shapes the Animal World—and Us*. Anchor Books. New York.

Ramirez, Steve, Christopher J. MacDonald, Anthony Moffa, Joanne Zhou, Roger L. Redondo, and Susumu Tonegawa (2015). "Activating positive memories engrams suppresses depression-like behaviour." *Nature*, *522*(7556), 335–9.

Raven, Catherine (2021). *Fox and I. An Uncommon Friendship*. Spiegel and Grau. New York.

Regan, Tom (2004). *The Case for Animal Rights*. University of California Press. Berkeley & Los Angeles, CA.

Regier, Jerome C., Jeffrey W. Shultz, Andreas Zwick, April Hussey, Bernard Ball, Regina Wetzer, Joel W. Martin, and Clifford W. Cunningham (2010). "Arthropod relationships revealed by phylogenomic analysis of nuclear protein- coding sequences." *Nature*, *463*(7284), 1079–83.

Regier, Jerome C., Charles Mitter, Andreas Zwick, Adam L. Bazinet, Michael P. Cummings, Akito Y. Kawahara, Jae-Cheon Sohn, Derrick J. Zwickl, Soowon Cho, Donald R. Davis, Joaquin Baixeras, John Brown, Cynthia Parr, Susan Weller, David C. Lees, and Kim T. Mitter (2013). "A large-scale, higher-level molecular phylogenetic study of the insect order Lepidoptera (moths and butterflies)." *PLoS ONE*, *8*(3), e58568.

Reindl, E., S.R. Beck, I.A. Apperly, and C. Tennie (2016). "Young children spontaneously invent wild great apes' tool-use behaviours." *Proc. Royal Soc. B*, *283*(1825), 20152402.

Reinhold, Annika Stefanie, Juan Ignacio Sanguinetti-Scheck, Konstantin Hartmann, and Michael Brecht (2019). "Behavioral and neural correlates of hide-and-seek in rats." *Science*, *365*(6458), 1180–3.

Rentmeester, Casey (2016). *Heidegger and the Environment*. Rowman and Littlefield. London & New York.

Rilke, Rainer Marie (2000). *Duino Elegies* (E. Snow, Trans.). North Point Press. New York.

Rosenberg, Kenneth V., Adriaan M. Dokter, Peter J. Blancher, John R. Sauer, Adam C. Smith, Paul A. Smith, Jessica C. Stanton, Arvind Panjabi, Laura Helft, Michael Parr, and Peter P. Marra (2019). "Decline of the North American Avifauna." *Science*, *366*(6461), 120–4.

Rus, Daniela, and Gregory Mone (2024). "The mind's mirror. Machine intelligences may teach us what it means to be human." *Natural History Magazine*, *132*(7), 28–33.

Rutz, Christian, Barbara C. Klump, Lisa Komarczyk, Rosanna Leighton, Joshua Kramer, Saskia Wischnewski, Shoko Sugasawa, Michael B. Morrissey, Richard James, James J.H. St Clair, Richard A. Switzer, and Bryce M. Masuda (2016). "Discovery of species-wide tool use in the Hawaiian crow." *Nature*, *537*(7620), 403–7.

Rygula, Rafal, Helena Pluta, and Piotr Popik (2012). "Laughing rats Are optimistic." *PLoS ONE*, *7*(12), e51959.

Safina, Carl (2020). *Becoming Wild. How Animal Cultures Raise Families, Create Beauty, and Achieve Peace*. Henry Holt. New York.

Safina, Carl (2023). *Alfie and Me. What Owls Know, What Humans Believe*. W.W. Norton. New York.

Sampaio, Eduardo, Vivek H. Sridhar, Fritz A. Francisco, Mátě Nagy, Ada Sacchi, Ariana Strandburg-Peshkin, Paul Nührenberg, Rui Rosa, Iain D. Couzin, and Simon Gingins (2024). "Multidimensional social influence drives leadership and composition-dependent success in octopus – fish hunting groups." *Nature Ecology & Evolution*, *8*(11), 2072–84.

Samuni, Liran, and Martin Surbeck (2023). "Cooperation across social borders in bonobos." *Science*, *382*(6672), 805–9.

Sanz, Crickette M., David Strait, Crepin Eyana Ayina, Jean Marie Assamba, Thierry Fabrice Ebombi, Severin Ndassoba Kialiema, Delon Ngoteni, Gaeton Mbebouti, David Rostand Koni Boue, Sean Brogan, Jake A. Funkhouser, and David B. Morgan (2022). "Interspecific interactions between sympatric apes." *iScience*, *25*(10), 105059.

Sato, Nobuya, Ling Tan, Kazushi Tate, and Maya Okada (2015). "Rats demonstrate helping behavior toward a soaked conspecific." *Animal Cognition*, *18*(5), 1039–47.

Scales, Helen (2018). *Eye of the Shoal. A Fishwatcher's Guide to Life, the Oceans and Everything* (Kindle ebook [www.amazon.com; downloaded June 18, 2018] ed.). Bloomsbury Sigma. London.

Schalow, Frank (2001). "At the crossroads of freedom: Ethics without values." *In*: R. Polt, and G. Fried (Eds.), *A Companion to Heidegger's Introduction to Metaphysics* (pp. 250–62). Yale University Press. New Haven, CT, & London.

Schalow, Frank (2006). *The Incarnality or Being. The Earth, Animals, and the Body in Heidegger's Thought*. State University of New York Press. Albany, NY.

Schalow, Frank (2020). *Historical Dictionary of Heidegger's Philosophy*. Rowman and Littlefield. Lanham, MD.

Scheel, David (2023). *Many Things Under a Rock. The Mysteries of Octopuses*. W.W. Norton. New York & London.

Scheel, David, Peter Godfrey-Smith, S. Linquist, S. Chancellor, M. Hing, and M. Lawrence (2018). "Octopus engineering, intentional and inadvertent." *Communicative and Integrative Biology*, *11*(1), e1395994.

Schloegl, Christian, Judith Schmidt, Markus Boeckle, Brigitte M. Weiß, and Kurt Kotrschal (2012). "Grey parrots use inferential reasoning based on acoustic cues alone." *Proc. Royal Soc. B*, *279*(1745), 4135–42.

Schmelz, Martin, Sebastian Grueneisen, Alihan Kabalak, Jürgen Jost, and Michael Tomasello (2017). "Chimpanzees return favors at a personal cost." *Proc. Natl. Acad. Sci. U.S.A.*, *114*(28), 7462–7.

Schötz, Susanne (2018). *The Secret Language of Cats. How to Understand Your Cat for a Better, Happier Relationship* (P. Kuras, Trans.). Hanover Square Press. New York.

Schultz, Ted R. (2020). "Fungus-farming ants (Attini in part)." *In*: Christopher Starr (Ed.), *Encyclopedia of Social Insects* (pp. 1–8). Springer International Publishing. Cham, Switzerland.

Schultz, Ted R. (2021). "The convergent evolution of agriculture in humans and fungus-farming ants." *In*: Ted R. Schultz, Richard Gawne, and Peter N. Peregrine (Eds.), *The Convergent Evolution of Agriculture in Humans and Insects* (pp. 281–313). The MIT Press. Cambridge, MA, & London.

Schultz, Ted R., Richard Gawne, and Peter N. Peregrine (Eds.). (2021). *The Convergent Evolution of Agriculture in Humans and Insects*. The MIT Press. Cambridge, MA, & London.

Schuster, Stefan (2018). "Hunting in archerfish – an ecological perspective on a remarkable combination of skills." *Journal of Experimental Biology, 221*(24), jeb159723.

Seeley, Thomas D. (1996). *The Wisdom of the Hive*. Harvard University Press. Cambridge, MA.

Seeley, Thomas D. (2024). *Piping Hot Bees and Boisterous Buzz-Runners*. Princeton University Press. Princeton, NJ, & Oxford.

Seth, Anil K., Bernard J. Baars, and David B. Edelman (2005). "Criteria for consciousness in humans and other Animals." *Consciousness and Cognition, 14*(1), 119–39.

Shubin, Neil (2009). *Your Inner Fish. A Journey into the 3.5-Billion-Year History of the Human Body*. Vintage Books. New York.

Singer, Peter (2002). *Animal Liberation*. HarperCollins. New York.

Singer, Peter (2023). *Animal Liberation Now. The Definitive Classic Renewed*. Harper Perennial. New York.

Sloutsky, Vladmir M., and Brandom M. Turner (2023). "Cognition: The power of simple associative learning." *Current Biology, 33*, R219–R237.

Smith, Jennifer E., and Mark Van Vugt (2020). "Leadership and status in mammalian societies: Context matters." *Trends in Cognitive Sciences, 24*(4), 263–4.

Smith, Jennifer E., Barbara Natterson-Horowitz, Maddison M. Mueller, and Michael E. Alfaro (2023). "Mechanisms of equality and inequality in mammalian societies." *Phil. Trans. Royal Soc. B, 378*(1883), 20220307.

Smith, John Maynard, and Eörs Szathmáry (2000). *The Origins of Life. From the Birth of Life to the Origins of Language*. Oxford University Press. New York.

Smith, N. Adam, Krista L. Koeller, Julia A. Clarke, Daniel T. Ksepka, Jonathan S. Mitchell, Ali Nabavizadeh, Ryan C. Ridgley, and Lawrence M. Witmer (2022). "Convergent evolution in dippers (Aves, Cinclidae): The only wing-propelled diving songbirds." *The Anatomical Record, 305*(7), 1563–91.

Sneddon, Lynne U., Victoria A. Braithwaite, and Michael J. Gentle (2003). "Novel object test: Examining nociception and fear in the rainbow trout." *The Journal of Pain, 4*(8), 431–40.

Snowdon, Charles T. (2017). "Learning from monkey 'talk.'" *Science, 355*(6330), 1120–2.

Staaf, Danna (2023). *The Lives of Octopuses and Their Relatives. A Natural History of Cephalopods*. Princeton University Press. Princeton, NJ, & Oxford.

Storey, David E. (2015). *Naturalizing Heidegger. His Confrontation with Nietzsche, His Contributions to Environmental Philosophy*. State University of New York. Albany, NY.

Stork, Nigel E. (2018). "How many species of insects and other arthropods are there on earth?" *Annual Review of Entomology, 63*, 31–45.

Swift, Kaeli N., and John M. Marzluff (2015). "Wild American crows gather around their dead to learn about danger." *Animal Behaviour, 109*, 187–97.

Taylor, Alex H., Rachael Miller, and Russell D. Gray (2012). "New Caledonian crows reason about hidden causal agents." *Proc. Natl. Acad. Sci. U.S.A., 109*(40), 16389–91.

Taylor, Alex H., Douglas M. Elliffe, Gavin R. Hunt, Nathan J. Emery, Nicola S. Clayton, and Russell D. Gray (2011). "New Caledonian crows learn the functional Properties of novel tool types." *PLoS ONE, 6*(12), e26887.

Thompson, Evan (2007). *Mind in Life. Biology, Phenomenology, and the Sciences of Mind*. Harvard University Press. Cambridge, MA, & London.

Toledo, Sivan, David Shohami, Ingo Schiffner, Emmanuel Lourie, Yotam Orchan, Yoav Bartan, and Ran Nathan (2020). "Cognitive map-based navigation in wild bats revealed by a new high-throughput tracking system." *Science*, *369*(6500), 188–93.

Tricarico, Elena, Luciana Borrelli, Francesca Gherardi, and Graziano Fiorito (2011). "I know my neighbour: Individual recognition in *Octopus vulgaris*." *PLoS ONE*, *6*(4), e18710.

Úbeda, Yulán, Sara Ortìn, Judy St. Leger, Miquel Llorente, and Javier Almunia (2019). Personality in captive killer whales (Orcinus orca): "A rating approach based on the five-factor model." *Journal of Comparative Psychology*, *133*(2), 252–61.

Vail, Alexander L., Andrea Manica, and Redouan Bshary (2013). "Referential gestures in fish collaborative hunting." *Nature Communications*, *4*(1), 1765.

Vail, Alexander L., Andrea Manica, and Redouan Bshary (2014). "Fish choose appropriately when and with whom to collaborate." *Current Biology*, *24*(17), R791–R793.

van Schaik, Carel P., Marc Ancrenaz, Gwendolyn Borgen, Biruite Galdikas, Cheryl D. Knott, Ian Singleton, Akira Suzuki, Sri Suci Utami, and Michelle Merrill (2003). "Orangutan cultures and the evolution of material culture." *Science*, *299*(5603), 102–5.

Van Vugt, Mark, and Jennifer E. Smith (2019). "A dual model of leadership and hierarchy: Evolutionary synthesis." *Trends in Cognitive Sciences*, *23*(11), 952–67.

Visalberghi, Elisabetta, Giulia Sirianni, Dorothy Fragaszy, and Christophe Boesch (2015). "Percussive tool use by Taï Western chimpanzees and Fazenda Boa Vista bearded capuchin monkeys: A comparison." *Phil. Trans. R. Soc. B*, *370*(1682), 20140351.

Vogel, Gretchen (1999). "Chimps in the wild show stirrings of culture." *Science*, *284*(5423), 2070–3.

von Bayern, A.M.P., S. Danel, A.M.I. Auersperg, B. Mioduszewska, and A. Kacelnik (2018). "Compound tool construction by New Caledonian crows." *Scientific Reports*, *8*(1), 15676.

von Uexküll, Jakob Johann (1982). "The theory of meaning." *Semiotica*, *42*(1), 25–82.

von Uexküll, Jakob Johann (2010). *A Foray into the Worlds of Animals and Humans, with a Theory of Meaning* (Joseph D. O'Neil, Trans.). University of Minnesota Press. Minneapolis, MN, & London.

von Uexküll, Jakob Johann (2018). *Theoretical Biology* (D.L. MacKinnon, Trans.). Forgotten Books. London.

Wang, Connie X., Isaac A. Hilburn, Daw-An Wu, Yuki Mizuhara, Christopher P. Cousté, Jacob N.H. Abrahams, Sam E. Bernstein, Avumu Matani, Shinsuke Shimoio, and Joseph L. Kirschvink (2019). "Transduction of the geomagnetic field as evidence from alpha-band activity in the human brain." *eNeuro*, *6*(2), e0483–18.

Wascher, Claudia A., Isabella B.R. Scheiber, and Kut Kotrschal (2008). "Heart rate modulation in bystanding geese watching social and non-social events." *Proc. Royal Soc. B*, *275*(1643), 1653–9.

Wasserman, Edward A., Andrew G. Kain, and Ellen M. O'Donoghue (2023). "Resolving the associative learning paradox by category learning in pigeons." *Current Biology*, *33*(6), 1112–16.

Wei-Chun, Ma, and David L. Denlinger (1974). "Secretory discharge and microflora of milk gland in tsetse flies." *Nature*, *247*(5439), 301–3.

Weir, Alex A.S., Jackie Chappell, and Alex Kacelnik (2002). "Shaping of hooks in New Caledonian Crows." *Science*, *297*(5583), 981.

Wills, Tom J., Francesca Cacucci, Neil Burgess, and John O'Keefe (2010). "Development of the hippocampal cognitive map in preweanling rats." *Science*, *328*(5985), 1573–80.

Wilson, David Sloan (2015). *Does Altruism Exist? Culture, Genes, and the Welfare of Others*. Yale University Press. New Haven, CT, & London.
Wilson, Edmund B. (1896). *The Cell in Development and Inheritance*. The Macmillan Company. New York & London.
Wilson, Edward O. (2012). *The Social Conquest of Earth*. Liveright Publishing. New York & London.
Wilson, Edward O. (2019). *Naturalist*. Island Press. Washington, DC.
Withy, Katherine (2013). "The strategic unity of Heidegger's *The Fundamental Concepts of Metaphysics*." *The Southern Journal of Philosophy*, *51*(2), 161–78.
Withy, Katherine (2015). *Heidegger on Being Uncanny*. Harvard University Press. Cambridge, MA, & London.
Withy, Katherine (2017). "Haugeland's Heidegger and the metaphysics of normativity." *European Journal of Philosophy*, *25*(2), 463–84.
Withy, Katherine (2022). *Heidegger on Being Self-concealing*. Oxford University Press. Oxford.
Wittek, Neslihan, Hiroshi Matsui, Nicole Kessel, Fatma Oeksuez, Onur Güntürkün, and Patrick Anselme (2021). "Mirror self- recognition in pigeons: Beyond the pass-or-fail criterion." *Frontiers in Psychology*, *12*, 669039.
Wolfe, Cary (2013). *Before the Law. Humans and Other Animals in a Biopolitical Frame*. The University of Chicago Press. Chicago, IL, & London.
Wong, Michael L.,Carol E. Cleland, Daniel Arend Jr., Stuart Bartlett, H. James Cleaves 2nd, Heather Demarest, Anirudh Prabhu, Jonathan I. Lunine, and Robert M. Hazen (2023). "On the roles of function and selection in evolving systems." *Proc. Natl. Acad. Sci. U.S.A.*, *120*(43), e2310223120.
Wrathall, Mark A. (Ed.). (2021). *The Cambridge Heidegger Lexicon*. Cambridge University Press. Cambridge & London.
Yasumuro, Haruhiko, and Yuzuru Ikeda (2018). "Environmental enrichment affects the ontogeny of learning, memory, and depth perception of the pharaoh cuttlefish *Sepia pharaonis*." *Zoology*, *128*, 27–37.
Yong, Ed (2022). *An Immense World. How Animal Senses Reveal the Hidden Realms Around Us*. Random House. New York.
Zhang, Shichang, Yangjié Liu, Yubing Ma, Hao Wang, Yao Zhao, Matjaž Kuntner, and Daiqin Li (2022). "Male spiders avoid sexual cannibalism with a catapult mechanism." *Current Biology*, *32*(8), R341–R359.
Zhao, Wenchan, Francisco Garcia-Oscos, Daniel Dinh, and Todd F. Roberts (2019). "Inception of memories that guide vocal learning in the songbird." *Science*, *366*(6461), 83–9.
Zimmerman, Michael E. (1993). "Rethinking the Heidegger—deep ecology relationship." *Environmental Ethics*, *15*(3), 195–224.
Zimmerman, Michael E. (2003). "Heidegger's Phenomenology and Contemporary Environmentalism." *In*: Charles S. Brown, and Ted Toadvine (Eds.), *Eco-Phenomenology. Back to the Earth Itself* (pp. 73–101). State University of New York Press.

INDEX

www.ingramcontent.com/pod-product-compliance
Ingram Content Group UK Ltd.
Pitfield, Milton Keynes, MK11 3LW, UK
UKHW020806290926
15470UKWH00003B/978

9 781350 604131